ELECTRICAL AND MECHANICAL OSCILLATIONS

An Introduction

BY

D. S. JONES

ROUTLEDGE AND KEGAN PAUL
LONDON

First published 1961
by Routledge & Kegan Paul Ltd
Broadway House, 68-74 Carter Lane
London, E.C.4

Printed in Great Britain
by Latimer, Trend & Co Ltd, Plymouth

ELECTRICAL AND MECHANICAL OSCILLATIONS

LIBRARY OF MATHEMATICS

edited by

WALTER LEDERMANN

D.S., Ph.D., F.R.S.Ed., Senior Lecturer in Mathematics, University of Manchester

Linear Equations	P. M. Cohn
Sequences and Series	J. A. Green
Differential Calculus	P. J. Hilton
Elementary Differential Equations and Operators	G. E. H. Reuter
Partial Derivatives	P. J. Hilton
Complex Numbers	W. Ledermann
Principles of Dynamics	M. B. Glauert
Electrical and Mechanical Oscillations	D. S. Jones
Vibrating Systems	R. F. Chisnell
Vibrating Strings	D. R. Bland
Fourier Series	I. N. Sneddon
Solutions of Laplace's Equation	D. R. Bland
Solid Geometry	P. M. Cohn

Contents

Preface

THIS book is concerned with the oscillations of electrical and mechanical systems. Only systems with one degree of freedom are discussed; other books in the series deal with the problems arising when a system has more than one degree of freedom or is continuous. Mechanical systems are considered first since this is the common practice in most courses. After a discussion of equilibrium positions of systems without friction there is an account of their oscillations near equilibrium. Next, the effects of friction of the viscous type are examined.

A separate chapter is devoted to electrical circuits so that those who wish to divorce them from mechanical systems may do so. However, the detailed mathematical analysis of earlier chapters has not been repeated; instead the results have been quoted. The ideas of complex and operational impedances which are more appropriate to electrical circuits than mechanical systems are also introduced in this chapter. Those who wish to develop the electrical theory at the same time as the mechanical will find some suitable examples at the end of Chapter Four.

The last chapter analyses oscillations which are not necessarily small in amplitude. The non-linearities of the system must then be taken into account. A strict mathematical approach to this subject is beyond the scope of this book but it is hoped that the plausible approach adopted will enable the student to grasp the essentials of the phenomena without becoming lost in mathematical rigour. Most students will undoubtedly find this the most difficult chapter—it is written in a more condensed style than the preceding chapters—but I hope they will find the problems so fascinating and of such practical importance as to be encouraged to study the more advanced works quoted.

I have endeavoured to make the examples bear some relation to practical reality although some of the simplifications necessary to make the mathematics tractable may appear artificial in the eyes of the reader. I have not always chosen units which are used in practice but have preferred those which arise naturally in the mathematical equations. In this way the confusion which occurs with units (e.g. whether or not the mass unit has to be operated on with g) should be avoided and the student's difficulties should lie only in the mathematics.

Finally, I wish to acknowledge with deep gratitude the assistance provided by Miss V. M. Cook who typed the manuscript both efficiently and elegantly.

D. S. Jones

University College of
North Staffordshire

CHAPTER ONE

Equilibrium

1.1. Introduction. Most of the oscillatory systems which occur in everyday life are more complicated than the ones which are discussed in this book. But we cannot hope to deal with the problems of rotating machinery, internal combustion engines, shock absorbers, aeroelasticity (to mention a few examples) until we have thoroughly understood the simplest oscillating systems we can imagine. Admittedly, the models we consider are only crude approximations to what we are likely to encounter in the world but they do demonstrate the main features that will be met.

It is important to realize that we are concerned with models of the physical world and that, as a consequence, some of the problems will appear highly artificial from a practical standpoint. For example, in this book, a spring is regarded as weightless, as always obeying Hooke's law and as having zero cross-sectional area so that it always lies along a straight line. Our reason for using such an abstraction instead of a real spring is that we believe it reproduces the main phenomena of oscillatory motion involving a spring while reducing considerably the mathematical complexity. The art of choosing the appropriate model is only acquired by experience which is one reason why science and its practical applications are often found difficult.

The theory of the motion of bodies is based upon certain ideas and laws which are considered in M. B. Glauert's book *Principles of Dynamics*, in this series. The fundamental law concerns the *particle*, or mass concentrated at a point. Let m be the mass of the particle and suppose that the particle is acted on by a force represented by the vector $\mathbf{F}$.

Let the position of the particle at the time t be specified by the position vector $\mathbf{r}$. Then the connection between force and motion is given by

$$\mathbf{F}=m\frac{d^2\mathbf{r}}{dt^2}. \tag{1}$$

A dot over a letter will usually be employed to indicate a derivative with respect to t. Thus $\dot{\mathbf{r}}$ and $\ddot{\mathbf{r}}$ signify $d\mathbf{r}/dt$ and $d^2\mathbf{r}/dt^2$ respectively. In this notation (1) can be written as

$$\mathbf{F}=m\ddot{\mathbf{r}}. \tag{2}$$

Suppose now that in the time δt the particle moves from the point $\mathbf{r}$ to the point $\mathbf{r}+\delta\mathbf{r}$. Then δW, *the work done by the force*, is defined by

$$\delta W=\mathbf{F}\cdot\delta\mathbf{r} \tag{3}$$

i.e. the scalar product of the force and displacement. If $\mathbf{F}$ and $\delta\mathbf{r}$ have components (X, Y, Z) and $(\delta x, \delta y, \delta z)$ respectively parallel to the Cartesian co-ordinate axes

$$\delta W=\mathrm{X}\delta x+\mathrm{Y}\delta y+\mathrm{Z}\delta z.$$

If we divide (3) by δt and proceed to the limit as $\delta t\to 0$ we see that dW/dt, the rate at which work is done, is given by

$$\dot{W}=\mathbf{F}\cdot\dot{\mathbf{r}}. \tag{4}$$

Another important quantity is the *kinetic energy* T which is defined as one-half the product of the mass and the square of its velocity, i.e.

$$T=\tfrac{1}{2}m\dot{\mathbf{r}}^2. \tag{5}$$

Substitute for $\mathbf{F}$ in (4) from (2); then

$$\dot{W}=m\ddot{\mathbf{r}}\cdot\dot{\mathbf{r}}=\dot{T} \tag{6}$$

from (5). Hence, *the rate of change of kinetic energy of the particle is equal to the rate at which work is done by the force applied to it.*

Let us assume that during the motion of the particle it moves from $\mathbf{r}_0$ to $\mathbf{r}_1$. The work done during this motion is, after integration of (3),

$$W = \int_{\mathbf{r}_0}^{\mathbf{r}_1} \mathbf{F} \cdot d\mathbf{r}.$$

In general, W will depend upon the path followed from $\mathbf{r}_0$ to $\mathbf{r}_1$ and different values will be obtained for different paths. It may happen, however, that W depends only on $\mathbf{r}_0$ and $\mathbf{r}_1$ and that the path between them is irrelevant. The force is then said to be *conservative* and we define the *potential energy* V by $V = -W$. Integration of (6) then gives

$$T + V = \text{constant},$$

i.e. for a conservative force, *the sum of the kinetic and potential energies is constant.*

All the preceding ideas can be generalized at once to a system containing n particles. Let the masses and positions of the particles be $m_1, m_2, \ldots, m_n$ and $\mathbf{r}_1, \mathbf{r}_2, \ldots, \mathbf{r}_n$ respectively. Then

$$\mathbf{F}_i = m_i \ddot{\mathbf{r}}_i \quad (i = 1, 2, \ldots, n)$$

where $\mathbf{F}_i$ is the total force acting on the ith particle. It includes not only the *external forces* originating from sources outside the system but also the *internal forces* (whether of electrical, gravitational or other nature) due to the other particles of the system.

The work done is defined as the sum of the separate contributions from each of the particles so that

$$\delta W = \sum_{i=1}^{n} \mathbf{F}_i \cdot \delta \mathbf{r}_i.$$

The kinetic energy T of the system is also defined as the sum of the individual kinetic energies of the particles; thus

$$T = \sum_{i=1}^{n} \tfrac{1}{2} m_i \dot{\mathbf{r}}_i^2. \tag{7}$$

In an analogous manner to that used in deriving (6) it can be proved that

$$\dot{W}=\dot{T}.$$

If *all* the forces, both internal and external, are conservative the system is said to be conservative. In a conservative system a potential energy $V=-W$ can be defined and

$$T+V=\text{constant},$$

i.e. *in a conservative system the sum of the kinetic and potential energies is constant.*

It should be remarked that even if the forces are not strictly conservative, the sum of the kinetic and potential energies may be constant. This will happen if the forces acting consist partly of conservative forces and partly of forces which do no work. Since the forces which do no work do not contribute to the energy equation this equation takes exactly the same form as if all the forces were conservative. In general, the forces which combine together to do no work can be excluded and the potential energy of the system calculated as the sum of the remaining separate potential energies. Thus the contributions of certain internal forces will cancel and they can be ignored in the system as a whole. For example, (1) the internal forces which keep two particles a constant distance apart whether they be gravitational or due to the tension in a taut string or rigid rod, (2) the mutual forces between two bodies smoothly pivoted together.

Two of the most important conservative forces from our point of view are gravity and the tension of a spring. We consider only local gravity which produces a constant acceleration g vertically downwards. The potential energy of a particle at height z is then mgz. As regards the *spring* we imagine it to be weightless and to have a uniform *tension* throughout its length. It is represented diagrammatically by a wiggly line as shown in Fig. 1.1. The *natural length* a of the spring is defined as the length at which the tension is zero. It will be assumed that the spring obeys Hooke's law of elasticity that the tension is proportional to the

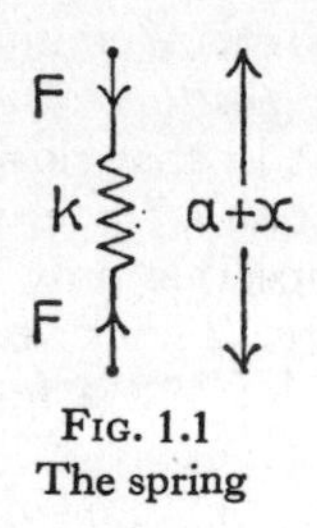

FIG. 1.1
The spring

extension. Then, when the length of the spring is $a+x$, the tension F is given by

$$F=kx. \tag{8}$$

The constant k is known as the *stiffness* of the spring. F is positive when x is positive and negative when x is negative; this displays the tendency of the spring to return to its natural length whether it be stretched or compressed.

The tension in an *elastic string* is also given by (8) when x is positive, ka then being known as the *modulus of elasticity*, i.e. tension=modulus×extension/original length, the modulus being independent of the string length. But, when x is negative, the force in the string is zero since the string cannot support compression. If k be large it is still possible for the tension to assume moderate values provided that x is small. We can visualize an extreme case in which k tends to infinity while x tends to zero in such a way that kx remains finite and non-zero. Idealizing in this way we obtain an *inextensible* string which can support tension without change of length.

The work done by the tension when the length of a spring is changed from a to $a+x$ is

$$\int_0^x -k\xi\, d\xi = -\tfrac{1}{2}kx^2.$$

This shows that the spring force is conservative with potential energy $\frac{1}{2}kx^2$.

1.2. The equilibrium of a conservative system with

one degree of freedom. A configuration of a dynamical system is said to be a *position of equilibrium* if, when the system is placed at rest in that position, it remains at rest. Since a particle subject to a force will accelerate according to (2) equilibrium is possible only if the total force acting on each particle is zero. This criterion is rather cumbersome to apply in practice and a more convenient one will now be derived.

In the first place we shall consider only those conservative systems whose position is completely specified as soon as one quantity q is known, i.e. systems of one degree of freedom. The position vector of any particle will then be a function of q alone. Consequently

$$\dot{\mathbf{r}}_i=\frac{d\mathbf{r}_i}{dt}=\frac{d\mathbf{r}_i}{dq}\frac{dq}{dt}=\frac{d\mathbf{r}_i}{dq}\dot{q}.$$

Therefore, formula (7) for the kinetic energy becomes

$$T=\tfrac{1}{2}\sum_{i=1}^{n} m_i\left(\frac{d\mathbf{r}_i}{dq}\right)^2\dot{q}^2=\tfrac{1}{2}M(q)\dot{q}^2$$

where $M(q)=\sum_{i=1}^{n} m_i\left(\frac{d\mathbf{r}_i}{dq}\right)^2$. M is a function of q only since each $\mathbf{r}_i$ is such a function. Furthermore, the kinetic energy can never be negative so that $M(q)>0$ whatever the value of q.

Since the system is conservative $T+V$ is constant, i.e.

$$\tfrac{1}{2}M(q)\dot{q}^2+V=\text{constant}. \qquad (9)$$

Now V involves only $\mathbf{r}_i, \ldots, \mathbf{r}_n$ and so is a function of q only. Consequently, if we take a time derivative of (9) and cancel a common factor† $\dot{q}$, we obtain

$$M\ddot{q}+\tfrac{1}{2}\frac{dM}{dq}\dot{q}^2+\frac{dV}{dq}=0. \qquad (10)$$

† This is permissible since the equation is valid as for all $\dot{q}$.

Let $q=\alpha$ be a position of equilibrium. Then when $q=\alpha$ and $\dot{q}=0$ there must be no acceleration, i.e. $\ddot{q}=0$. It follows from (10) that we must have

$$\left(\frac{dV}{dq}\right)_{q=\alpha}=0. \tag{11}$$

Conversely, suppose that (11) is true. Then (10) shows that when $q=\alpha$ and $\dot{q}=0$ we must have $\ddot{q}=0$ since $M\neq 0$. But this is not enough to show that the system does not move for the third or higher derivative of q might be non-zero at $q=\alpha$. However, a time derivative of (10) gives

$$M\dddot{q}+2\frac{dM}{dq}\dot{q}\,\ddot{q}+\tfrac{1}{2}\frac{d^2M}{dq^2}\dot{q}^3+\frac{d^2V}{dq^2}\dot{q}=0$$

which shows that $\dddot{q}$ vanishes when both $\dot{q}$ and $\ddot{q}$ do. Taking another time derivative we see that the fourth derivative is zero when the preceding ones are and, continuing the process, we can see that all derivatives of q vanish so that it must maintain a constant value. The system is therefore in equilibrium.

What we have demonstrated is that a *necessary and sufficient condition for* $q=\alpha$ *to be a position of equilibrium for a conservative system of one degree of freedom is* $\left(\frac{dV}{dq}\right)_{q=\alpha}=0.$

It should be remarked that the proof of sufficiency involves the assumption that q can be expanded in a power series near $q=\alpha$ and that finite derivatives of all orders of M and V exist. This is true in most practical cases but there may be cases, e.g. $V=-\frac{9}{2}q^{4/3}$, $M=1$ where the conditions are not satisfied and the theorem fails.

Example 1.1. As our example we shall consider the simple problem of a particle of mass m suspended by a spring of stiffness k from a fixed platform (Fig. 1.2). Both the spring tension and the force of gravity are conservative so that the system is conservative. Let the length of the spring be $a+q$, a being the

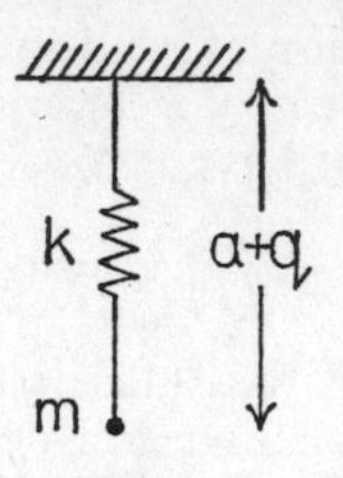

FIG. 1.2
Mass suspended by a spring

natural length. Then the potential energy of the spring is $\frac{1}{2}kq^2$ and that of the mass is $-mg(a+q)$. Hence

$$V=\tfrac{1}{2}kq^2-mg(a+q)$$

and

$$\frac{dV}{dq}=kq-mg.$$

Thus there is one position of equilibrium given by

$$q=mg/k.$$

In this position the spring tension exactly balances the force of gravity so that there is no net force on the particle.

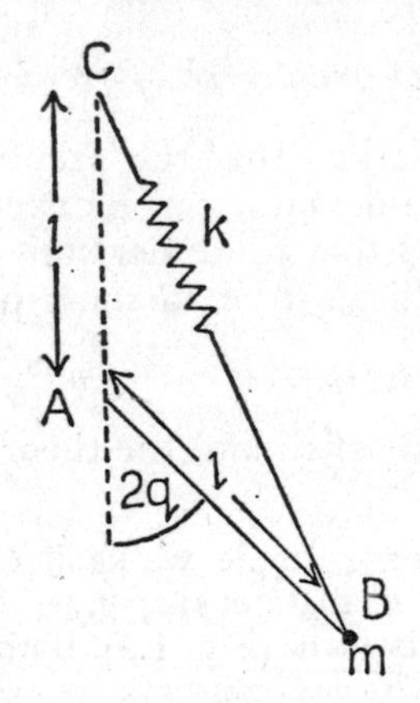

FIG. 1.3

Example 1.2. A light rigid rod AB is smoothly pivoted at A and carries a particle of mass m at B. A spring of natural length a and stiffness k is connected between B and C, a point vertically above A such that $AB = AC = l$ (Fig. 1.3).

The forces of the smooth pivots at A and C do no work; nor do the mutual reactions between the spring and rod at B. The spring tension and force of gravity are conservative. Therefore the system is conservative.

Let AB make an angle $2q$ with the downward vertical. ABC is an isosceles triangle with $B\hat{A}C = \pi - 2q$. Therefore $BC = 2\,l \cos q$ and the potential energy of the spring is $\frac{1}{2}k(2l \cos q - a)^2$. The depth of B below A is $l \cos 2q$ so that the potential energy of the mass is $-mgl \cos 2q$.

Hence

$$V = \tfrac{1}{2}k(2l \cos q - a)^2 - mgl \cos 2q$$

and

$$\frac{dV}{dq} = -2k\,l \sin q\,(2l \cos q - a) + 2mg\,l \sin 2q.$$

Since $\sin 2q = 2 \sin q \cos q$ the positions of equilibrium are given by

$$2l \sin q\{2mg \cos q - k(2l \cos q - a)\} = 0.$$

Therefore, either $\sin q = 0$

or $$\cos q - \tfrac{1}{2}ka/(kl - mg). \tag{12}$$

The first equation supplies $q = 0$ since the largest value of q is $\frac{1}{2}\pi$. There is therefore a position of equilibrium with B vertically below A.

With regard to the second equation, $0 \leqslant \cos q \leqslant 1$ because $|q| \leqslant \frac{1}{2}\pi$ and so a value of q can be found only when $kl > mg$ and $\frac{1}{2}ka/(kl - mg) \leqslant 1$. When these inequalities are satisfied there are 2 positions of equilibrium symmetrically placed about the vertical.

To summarize, there is at least one position of equilibrium and there may be three.

1.3. Stability. A dynamical system in equilibrium will rarely be free from disturbances of one kind or another, e.g. that produced by a sudden gust of wind. The question therefore arises as to whether a system will remain near equilibrium when a slight impulse or slight displacement is applied. If, when the system is given a small kinetic energy and a small displacement from equilibrium, the subsequent kinetic energy and displacement are small the equilibrium position is called *stable*. Otherwise it is *unstable*.

Let the potential energy at the position of equilibrium be V_0. Suppose that the system is started with kinetic energy δT_0 from a position where the potential energy is $V_0+\delta V_0$. Then, in the subsequent motion,

$$T+V=\delta T_0+V_0+\delta V_0$$

or

$$T+v=\delta T_0+\delta V_0 \tag{13}$$

on putting $V=V_0+v$.

Now let V always increase as the departure from equilibrium increases, i.e. $v\geqslant 0$. Then, since $T\geqslant 0$, (13) shows that $0\leqslant v\leqslant \delta T_0+\delta V_0$. Thus v is always a small positive quantity and the system never moves very far from equilibrium. Furthermore, since $v\geqslant 0$, (13) gives $0\leqslant T\leqslant \delta T_0+\delta V_0$ which demonstrates that the kinetic energy remains small. The position of equilibrium is therefore stable.

If, however, $v\leqslant 0$ (13) shows that T does not decrease. But, if the system is to remain near equilibrium, $\dot{q}$ must pass through a zero and change sign otherwise the magnitude of q would increase without limit. Thus T must become zero at some point. Since this is impossible if $v\leqslant 0$ the position is unstable.

It may happen, of course, that $v>0$ for $q>0$ and $v\leqslant 0$ for $q<0$. In that case consideration of the side $q<0$ alone indicates that the position is unstable. Hence *a necessary and sufficient condition for a position of equilibrium to be stable is that the potential energy increases in displacements from equilibrium.*

Since we are concerned only with the immediate neighbourhood of a position of equilibrium we can ensure that the potential energy increases by requiring it to be a minimum at equilibrium. Thus $\left(\dfrac{d^2V}{dq^2}\right)_{q=\alpha}>0$ is sufficient to make a position of equilibrium stable. If $\left(\frac{d^2V}{dq^2}\right)_{q=\alpha}=0$ the position may still be stable, e.g. if $\left(\frac{d^3V}{dq^3}\right)_{q=\alpha}=0$ and $\left(\frac{d^4V}{dq^4}\right)_{q=\alpha}>0$.

The rule can be remembered easily by drawing the curve of V against q and imagining it to be a fine wire. If a small bead is placed on the wire it will remain at rest in the position of equilibrium. If the bead is slightly disturbed it will not move very far away in the case of Fig. 1.4 where the potential energy has a minimum so that the position is stable. In the cases of Fig. 1.5, however, the bead will

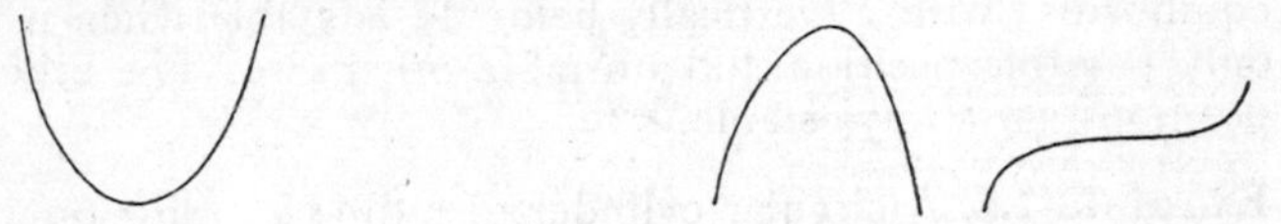

FIG. 1.4. Stable position FIG. 1.5. Unstable positions

obviously slide down the lower portions and the positions are unstable.

It is important to emphasize that we have discussed stability only with reference to an initial small disturbance. If impulses were arriving all the time they might be so synchronized that a large displacement was produced from an accumulation of small ones. Moreover, although small displacements cannot occur near an unstable position, there is nothing to prevent a motion which is not infinitesimal, e.g. an oscillation from A to B in Fig. 1.6 which passes through 2 stable and 1 unstable position of equilibrium.

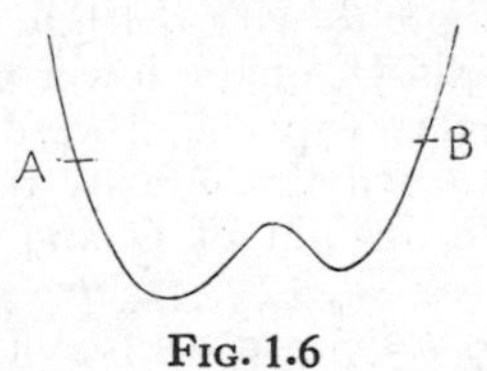

FIG. 1.6

Example 1.3. The stability of the system in Ex. 1.1.

Here $\dfrac{d^2V}{dq^2}=k>0$ at the position of equilibrium so that it is stable.

Example 1.4. The stability of the system in Ex. 1.2.

Here

$$\frac{d^2V}{dq^2}=2\,l\cos q\{2(mg-kl)\cos q+ka\}-4\,l(mg-kl)\sin^2 q. \quad (14)$$

At the position of equilibrium $q=0$ the right-hand side is $2l\{2(mg-kl)+ka\}$ which is positive or negative according as $\frac{1}{2}ka \gtrless kl-mg$. At the positions of equilibrium given by (12) only the last term of (14) survives and the derivative is positive or negative according as $kl \gtrless mg$. On account of the conditions for the existence of the side positions of equilibrium we can say that equilibrium with B vertically below A is stable when it is the only possible position and unstable otherwise. The other two positions are always stable.

Example 1.5. A circular cylinder of radius b rolling on a fixed circular cylinder of radius a, the centre of mass G of the upper cylinder being a distance c from its axis (Fig. 1.7).

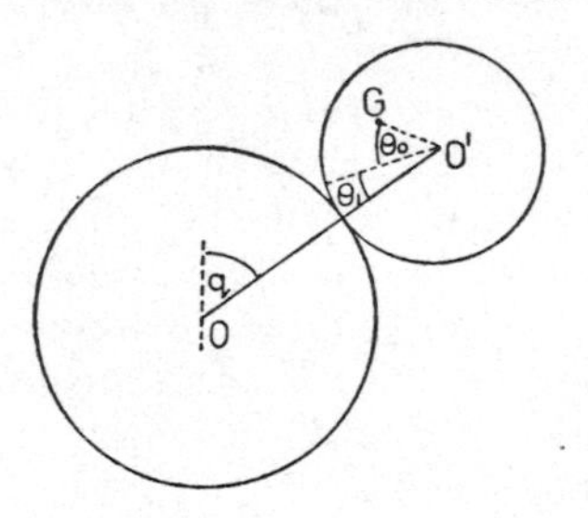

FIG. 1.7

Let OG make an angle θ_0 with the line of centres when O is vertically above the axis O' of the lower cylinder. Let the line of centres make an angle q with the upward vertical after rolling. Then the angle θ_1, turned through by the upper cylinder, is given by $aq=b\theta_1$. The height of G above O' is $(a+b)\cos q-c\cos(\theta_0+\theta_1+q)$. Hence

$$V=(a+b)\cos q-c\cos\{\theta_0+(1+a/b)q\}$$

and

$$\frac{dV}{dq} = -(a+b)\sin q + c(1+a/b)\sin\{\theta_0 + (1+a/b)q\}.$$

Thus the positions of equilibrium satisfy

$$b \sin q = c \sin\{\theta_0 + (1+a/b)q\}$$

which expresses the fact that G is vertically above the line of contact.

Also

$$\frac{d^2V}{dq^2} = -(a+b)\cos q + c(1+a/b)^2 \cos\{\theta_0 + (1+a/b)q\}.$$

At the position of equilibrium let h be the height of G above the line of contact, i.e. $h = b\cos q - c\cos\{\theta_0 + (1+a/b)q\}$. Then

$$\frac{d^2V}{dq^2} = \frac{a+b}{b^2}\{ab\cos q - h(a+b)\}$$

and the equilibrium is stable if

$$\frac{\cos q}{h} > \frac{1}{a} + \frac{1}{b}. \tag{15}$$

Note that if 2 cylinders, whose cross-sections are not necessarily circular, are in equilibrium the centre of mass of the upper is above the line of contact. In a small rolling motion the 2 cylinders would behave as 2 circular cylinders with radii equal to the radii of curvature at the line of contact. Therefore (15) would imply stability, a and b being the radii of curvature at the line of contact, q the inclination of the common tangent to the horizontal and h the height of the centre of mass above the line of contact.

1.4. The principle of virtual work. In a conservative system $T+V=$constant and therefore $d(T+V)/dt=0$. Now

$$\frac{dT}{dt} = \sum_{i=1}^{n} m_i \dot{\mathbf{r}}_i \cdot \ddot{\mathbf{r}}_i.$$

At a position of equilibrium all accelerations are zero. Hence $dT/dt=0$ at a position of equilibrium whatever the velocities. Consequently $dV/dt=0$ at a position of equilibrium.

This result is often known as the *principle of virtual work*—at a position of equilibrium $dV/dt=0$ for every possible set of velocities.

The implications of this principle are important. Consider, for example, the case when the whole system is specified by only one co-ordinate q. Then $\dfrac{dV}{dt}=\dfrac{dV}{dq}\dot{q}$ which can vanish for all possible $\dot{q}$ only if $dV/dq=0$. Thus we recover the equation already derived in 1.2 to determine the positions of equilibrium.

Suppose now that the system requires 2 co-ordinates q_1 and q_2 to specify it. Then

$$\frac{dV}{dt}=\frac{\partial V}{\partial q_1}\dot{q}_1+\frac{\partial V}{\partial q_2}\dot{q}_2.$$

At a position of equilibrium this must vanish for all $\dot{q}_1$ and $\dot{q}_2$. A possible selection is $\dot{q}_2=0$, $\dot{q}_1\neq 0$; then we must have

$$\frac{\partial V}{\partial q_1}=0. \tag{16}$$

On the other hand, a possible set of velocities is $\dot{q}_1=0$, $\dot{q}_2\neq 0$ so that

$$\frac{\partial V}{\partial q_2}=0. \tag{17}$$

Consequently, the positions of equilibrium are given by those values of q_1 and q_2 which satisfy *simultaneously* (16) and (17).

It will be observed that in both of the cases we have considered V is stationary at positions of equilibrium. Quite generally, the principle of virtual work shows that *the potential energy is stationary at a position of equilibrium* whatever number of co-ordinates is needed to specify the system.

As far as stability is concerned the argument of 1.3 showing that a position is stable if V increases away from equilibrium is independent of the number of co-ordinates. Moreover, if there is some direction in which V decreases we choose a co-ordinate q along this direction and consider motion in which only q varies. A repetition of the argument concerning instability reveals that the same is true here. Hence, *a necessary and sufficient condition for a position of equilibrium to be stable is that the potential energy increase in an arbitrary displacement from equilibrium.*

For example, in a system with 2 degrees of freedom and potential energy $V(q_1, q_2)$ let $q_1=\alpha_1$, $q_2=\alpha_2$ be a position of equilibrium. Then

$$\left(\frac{\partial V}{\partial q_1}\right)_\alpha=0, \quad \left(\frac{\partial V}{\partial q_2}\right)_\alpha=0 \tag{18}$$

where $(\quad)_\alpha$ means put $q_1=\alpha_1$ and $q_2=\alpha_2$. Put $q_1=\alpha_1+\epsilon_1$, $q_2=\alpha_2+\epsilon_2$ and expand in a Taylor series; we obtain

$$\begin{aligned} &V(\alpha_1+\epsilon_1,\ \alpha_2+\epsilon_2)-V(\alpha_1,\ \alpha_2) \\ &=\tfrac{1}{2}\epsilon_1^2\left(\frac{\partial^2 V}{\partial q_1^2}\right)_\alpha+\epsilon_1\epsilon_2\left(\frac{\partial^2 V}{\partial q_1 \partial q_2}\right)_\alpha+\tfrac{1}{2}\epsilon_2^2\left(\frac{\partial^2 V}{\partial q_2^2}\right)_\alpha \end{aligned}$$

because of (18), neglecting third-order terms. If the right-hand side is positive for all ϵ_1 and ϵ_2 it is positive when $\epsilon_1\neq 0$, $\epsilon_2=0$; this requires

$$\left(\frac{\partial^2 V}{\partial q_1^2}\right)_\alpha>0. \tag{19}$$

Knowing this we can rewrite the right-hand side as

$$\left[\left\{\left(\frac{\partial^2 V}{\partial q_1^2}\right)_\alpha\epsilon_1+\left(\frac{\partial^2 V}{\partial q_1 \partial q_2}\right)_\alpha\epsilon_2\right\}^2 + \left\{\left(\frac{\partial^2 V}{\partial q_1^2}\right)_\alpha\left(\frac{\partial^2 V}{\partial q_2^2}\right)_\alpha-\left(\frac{\partial^2 V}{\partial q_1 \partial q_2}\right)_\alpha^2\right\}\epsilon_2^2\right]\Big/ 2\left(\frac{\partial^2 V}{\partial q_1^2}\right)_\alpha$$

which will be positive for all ϵ_1 and ϵ_2 if, and only if,

$$\left(\frac{\partial^2 V}{\partial q_1^2}\right)_\alpha\left(\frac{\partial^2 V}{\partial q_2^2}\right)_\alpha>\left(\frac{\partial^2 V}{\partial q_1 \partial q_2}\right)_\alpha^2. \tag{20}$$

Conditions (19) and (20) therefore are sufficient to ensure stability in a system with 2 degrees of freedom.†

† The general theory of stationary points of functions of two variables is discussed by P. J. Hilton in *Partial Derivatives* in this series.

Example 1.6. A picture frame, of width $2ae$ ($e<1$) and depth $2h$, is hung by an inextensible string of length $2a$ attached to its upper corners and passing over a smooth peg. (Fig. 1.8.)

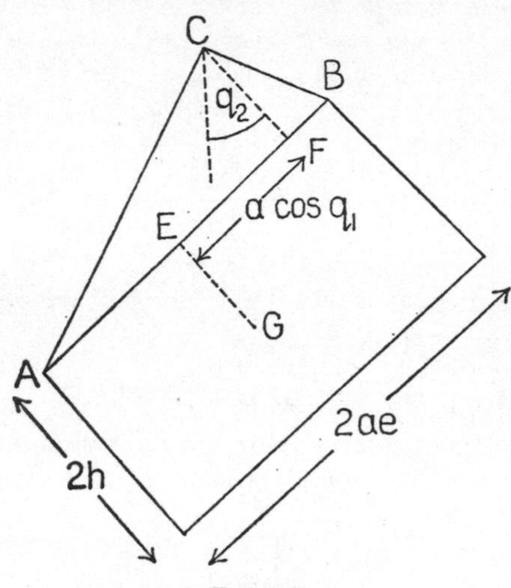

FIG. 1.8

The tension of the inextensible string does no work, neither does the reaction at the smooth peg nor do the mutual reactions at A and B. The system is conservative with gravity the only force contributing to the potential energy.

Let GE and CF be the perpendiculars to upper edge AB from the centre of mass G of the frame and peg C respectively. Let CF make an angle q_2 with the downward vertical and let $EF=a\cos q_1$ (this is possible since $EB=ae<a$). From triangles AFC and BFC

$$(ae+a\cos q_1)^2+CF^2=AC^2, \tag{21}$$
$$(ae-a\cos q_1)^2+CF^2=(2a-AC)^2.$$

By subtraction we obtain

$$AC=a(1+e\cos q_1)$$

and, after substitution in (21),

$$CF=a\sqrt{(1-e^2)}\sin q_1.$$

The depth of G below C is $(CF+EG)\cos q_2+EF\sin q_2$ and hence

$$V=-mg\{(b\sin q_1+h)\cos q_2+a\cos q_1\sin q_2\}$$

where $b=a\sqrt{(1-e^2)}$ and m is the mass of the frame. At a position of equilibrium V is stationary and

$$\frac{\partial V}{\partial q_1}=-mg\{b\cos q_1\cos q_2-a\sin q_1\sin q_2\}=0, \qquad (22)$$

$$\frac{\partial V}{\partial q_2}=-mg\{a\cos q_1\cos q_2-(b\sin q_1+h)\sin q_2\}=0. \qquad (23)$$

Subtract a/b times (22) from (23):

$$b(b\sin q_1+h)\sin q_2-a^2\sin q_1\sin q_2=0.$$

Therefore either $\sin q_2=0$ or $\sin q_1=bh/(a^2-b^2)=bh/a^2e^2$. If $\sin q_2=0$ then, from (22), $\cos q_1=0$ since $\cos q_2\neq 0$. Thus $EF=0$ and $AC=CB$. Thus, as might be expected, the symmetrical configuration with AB horizontal is a position of equilibrium.

The other possibilities exist only if $bh\leqslant a^2e^2$ and then $a\tan q_2=b\cot q_1$. These provide configurations on either side of the vertical with AB inclined to the horizontal.

With regard to stability it can be shown that, when $q_1=\frac{1}{2}\pi$, $q_2=0$ (corresponding to the symmetrical position with AB below C),

$$\frac{\partial^2 V}{\partial q_1^2}=mgb,\quad \frac{\partial^2 V}{\partial q_1\partial q_2}=mga,\quad \frac{\partial^2 V}{\partial q_2^2}=mg(h+b).$$

Consequently, (19) is automatically satisfied, whereas (20) holds if $hb>a^2e^2$. The central position is stable when there are no side positions and unstable otherwise.

It can also be proved, though a little algebraic manipulation is needed, that the side positions when they exist are stable.

The reader may be inclined to doubt that the symmetrical position can ever be unstable in practice, but he can easily confirm our conclusions by thinking about the extreme case in which $h=0$. If he wishes to conduct experiments he should be warned that it is difficult to find smooth picture cord. If the friction of the string is sufficient to prevent sliding then, to a first approximation, q_1 has the constant value $\frac{1}{2}\pi$ and for variations in q_2 *alone* the position is stable, i.e. instability is caused only by those disturbances which make the string slide.

EXERCISES ON CHAPTER I

1. A heavy uniform rod ACB of length $2a$ rests with A in contact with a smooth vertical wall and C on a smooth horizontal peg. If C is distant b from the wall find the inclination of the rod to the vertical.

2. A uniform rod ABC of length $2a$ and mass m is smoothly hinged at A. One end of a spring of natural length a and stiffness k is attached to B and the other end to D, a point vertically above A. If $AB = AD = \frac{3}{2}a$ find the positions of equilibrium and examine their stability.

3. A bead of mass m can slide on a smooth vertical circular wire of radius a. An inextensible string, which is attached to it, passes over a smooth peg at the highest point of the circle and carries a mass m_1 at the other end. Show that there is one equilibrium position if $2m < m_1$ but three if $2m > m_1$. Prove also that the symmetrical position is unstable when it is the only one but stable otherwise.

4. Four equal uniform rods of length a and mass m form a rhombus $ABCD$, being smoothly pivoted at A, B, C and D. An elastic string of modulus mg and natural length $a\sqrt{2}/3$ joins A and C. The system hangs in equilibrium from A. Find the inclination of the rods to the vertical.

5. A uniform plank of thickness $2h$ can roll on a fixed horizontal circular cylinder of radius a. If the plank is in equilibrium when it is horizontal prove that the position is stable or unstable according as $a \gtrless h$.

6. An inverted pendulum consists of a light rigid rod of length l carrying a mass m at its upper end and smoothly pivoted at its lower end. At a distance d along the rod from the pivot two horizontal springs, each of stiffness k, are attached and join the rod to two vertical walls on either side of it. Find the condition that the vertical position of equilibrium is stable.

7. A cylindrical body is in equilibrium on a plane inclined at an angle α to the horizontal. The generators of the cylinder are horizontal. Show that the equilibrium is stable if $h < R \cos \alpha$ where h is the height of the centre of mass above the line of

contact and R is the radius of curvature at the line of contact of the cylinder cross-section.

8. A uniform rod AB of length l hangs by two equal crossed inextensible strings CB and DA from two points C and D on the same horizontal level. The rod is restricted to the vertical plane through CD and $CD = l$, $CB = d$. Prove that the symmetrical position of equilibrium is stable if $l < \sqrt{(d^2 - l^2)}$.

9. Two equal uniform rods AB, BC of mass m and smoothly pivoted at B hang from a fixed smooth pivot at A. An inextensible string is connected to C, passes over a smooth pulley and carries a mass M at its other end. If the height of the pulley is automatically adjusted so that the portion of string at C is horizontal show that, in equilibrium, AB and BC are inclined to the vertical at angles $\tan^{-1}(2M/3m)$ and $\tan^{-1}(2M/m)$ respectively.

CHAPTER TWO

Harmonic Oscillations

2.1. The motion of a conservative system near equilibrium. In the preceding chapter the question as to whether a slightly disturbed conservative system remains near or departs from equilibrium was resolved but there was no attempt to discuss the actual form of the motion. We shall now examine the dynamical behaviour near equilibrium more closely. Our considerations will be limited to conservative systems with one degree of freedom.

According to (10) of Chapter I the equation of motion for a conservative system with one degree of freedom can be written

$$M\ddot{q}+\tfrac{1}{2}\frac{dM}{dq}\dot{q}^2+\frac{dV}{dq}=0 \tag{1}$$

where M and V are functions of q only. Let $q=\alpha$ be a position of equilibrium. Put $q=\alpha+x$ so that $\dot{q}=\dot{x}$ and $\ddot{q}=\ddot{x}$. At present we are interested only in motion near equilibrium so that it is plausible to assume that x, $\dot{x}$ and $\ddot{x}$ are all small quantities of the same order. (Attempts to prove this rigorously are by no means straightforward.) This suggests that we should expand in terms of the small quantities and reject all but the most important. For example,

$$M(\alpha+x)=M(\alpha)+\text{a quantity of the first order}$$

and therefore

$$M\ddot{q}=M(\alpha)\ddot{x}+\text{a quantity of the second order.}$$

Similarly,

$$\frac{dM}{dq}\dot{q}^2=\text{a quantity of the second order.}$$

Also, by Taylor's theorem,

$$\left(\frac{dV}{dq}\right)_{q=\alpha+x}=\left(\frac{dV}{dq}\right)_{q=\alpha}+x\left(\frac{d^2V}{dq^2}\right)_{q=\alpha}+\text{second order terms.}$$

The term $(dV/dq)_{q=\alpha}$ vanishes because $q=\alpha$ is a position of equilibrium. Hence, when we substitute our expansions in (1), we obtain

$$M(\alpha)\ddot{x}+\left(\frac{d^2V}{dq^2}\right)_{q=\alpha}x+\text{second order terms}=0.$$

Neglecting the second order terms in comparison with the first order terms we derive the following differential equation for x:

$$M(\alpha)\ddot{x}+\left(\frac{d^2V}{dq^2}\right)_{q=\alpha}x=0. \quad (2)$$

$M(\alpha)$ and $(d^2V/dq^2)_{q=\alpha}$ are constants. Consequently, the equation is a linear differential equation with constant coefficients. The theory of differential equations of this type will be found in the book *Elementary Differential Equations and Operators* by G. E. H. Reuter.† It is shown in **RI2.1** that the general solution of (2) is

$$x=A\,e^{kt}+B\,e^{-kt} \quad (3)$$

where A and B are arbitrary constants (possibly complex) and

$$M(\alpha)k^2+\left(\frac{d^2V}{dq^2}\right)_{q=\alpha}=0. \quad (4)$$

If $(d^2V/dq^2)_{q=\alpha}>0$, $k^2<0$ because $M>0$. Therefore k is purely imaginary and we write $k=i\Omega$ where Ω is real. The solution (3) can then be expressed as

$$x=C\cos\Omega t+D\sin\Omega t \quad (5)$$

where C and D are real constants and

$$M(\alpha)\Omega^2=(d^2V/dq^2)_{q=\alpha}.$$

The differential equation (2) can be written, in this notation, as

$$\ddot{x}+\Omega^2x=0. \quad (6)$$

† References to this book will be denoted by **R**, e.g. **RI2.1** means section 2.1 of Chapter I of Reuter's book.

The nature of the solution depends upon the sign of d^2V/dq^2. If the sign is negative k is real and, in any motion in which $A \neq 0$, x increases exponentially. We cannot conclude that this will happen indefinitely because eventually x and its derivatives will become large enough to violate the assumptions on which the derivation of (2) was based. Nevertheless we have another indication of the instability of a position of equilibrium where the potential energy is a maximum—there are certainly small disturbances for which the subsequent motion is not small.

Note that if $d^2V/dq^2=0$ the solution of (2) is not (3) but $x=A+Bt$. Again there is instability but the question arises immediately as to whether it is correct to treat $\ddot{x}$ and x as of the same order. This difficult problem we shall leave on one side.

At a stable position of equilibrium (5) is applicable. It can be cast into another form by introducing the angle ϵ defined by

$$\cos \epsilon = C/X, \quad \sin \epsilon = -D/X$$

where $X=\sqrt{(C^2+D^2)}$. Then

$$x = X \cos (\Omega t + \epsilon). \tag{7}$$

A motion in which (5) or (7) is valid is called a *simple harmonic motion*. The positive quantity X is called the

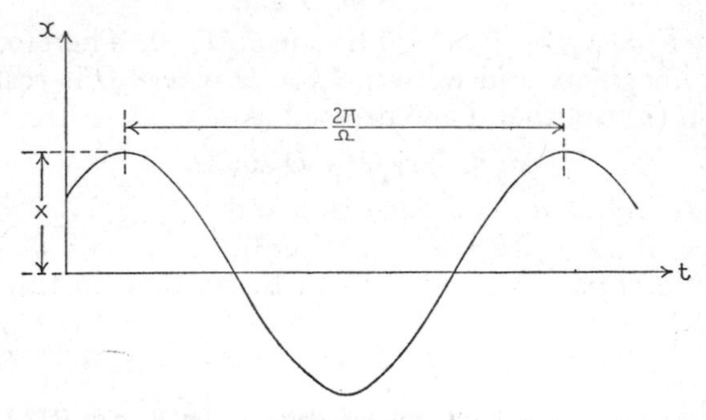

FIG. 2.1
Simple harmonic motion

amplitude because, as is easily seen from Fig. 2.1, $|x| \leqslant X$ at all times. If t is replaced by $t+2\pi/\Omega$ the values of x, $\dot{x}$ and $\ddot{x}$ are unaltered. For this reason $2\pi/\Omega$ is known as the *period*. The number of oscillations (or cycles) per second, or *frequency*, is $\Omega/2\pi$. Ω itself is often known as the *circular frequency*. The quantity $\Omega t+\epsilon$ is called the *phase* of the oscillation (some writers reserve this term for ϵ).

The reader should observe that, if X is small, so are x, $\dot{x}$ and $\ddot{x}$ because $|x| \leqslant X$, $|\dot{x}| \leqslant \Omega X$, $|\ddot{x}| \leqslant \Omega^2 X$. Therefore the assumptions on which we based our theory are satisfied by this solution. Hence, after a slight disturbance from stable equilibrium, *the system oscillates with simple harmonic motion about the position of equilibrium.*

The quantities X and ϵ are undetermined until further information is supplied. Usually initial conditions are the ones which arise. Let $x=x_0$ and $\dot{x}=x_1$ when $t=0$. From (5)

$$C=x_0, \quad \Omega D=x_1$$

and so

$$X=\surd(x_0^2+x_1^2/\Omega^2).$$

Hence

$$x=x_0 \cos \Omega t+\frac{x_1}{\Omega} \sin \Omega t$$
$$=\surd(x_0^2+x_1^2/\Omega^2) \cos (\Omega t+\epsilon)$$

where

$$\cos \epsilon=\frac{x_0}{\surd(x_0^2+x_1^2/\Omega^2)}, \quad \sin \epsilon=\frac{-x_1}{\surd(x_0^2\Omega^2+x_1^2)}.$$

2.2. Examples. Although the general theory tells us what must occur near a stable position of equilibrium it is not necessarily the best approach for dealing with particular problems. The fundamental equation which must be arrived at is (2) and there are two main methods of deriving it.

The first method rests on the observation that (2) is an equation for the acceleration in which the second and higher order terms have been neglected. Therefore, if all the equations of motion ($\mathbf{F}=m\ddot{\mathbf{r}}$) appropriate to the system are written down and second and higher order terms rejected, (2) should follow at once.

The aim of the second method is to derive (2) from the energy equation without the intermediate (and often laborious) step (1). This is achieved by carrying out the approximation one stage earlier than was done in section 2.1. To this end we expand T and V about the position of equilibrium and obtain

$$T=\tfrac{1}{2}M(q)\dot{q}^2=\tfrac{1}{2}M(\alpha)\dot{x}^2+\text{third order terms,}$$

$$V(q)=V(\alpha)+\tfrac{1}{2}\left(\frac{d^2V}{dq^2}\right)_{q=\alpha}x^2+\text{third order terms.}$$

The energy conservation equation gives

$$\tfrac{1}{2}M(\alpha)x^2+V(\alpha)+\tfrac{1}{2}\left(\frac{d^2V}{dq^2}\right)_{q=\alpha}x^2+\text{third order terms}$$
$$=\text{constant.}$$

On neglecting third and higher order terms the equation becomes

$$\tfrac{1}{2}M(\alpha)\dot{x}^2+V(\alpha)+\tfrac{1}{2}\left(\frac{d^2V}{dq^2}\right)_{q=\alpha}x^2=\text{constant.}$$

A derivative with respect to t recovers (2). Hence, there is the following rule for small oscillations near equilibrium: *calculate the kinetic and potential energies correct to the second order and then take a time derivative of the equation expressing conservation of energy.*

Either method must lead to the same result eventually. In some cases one is faster than the other but no general rule can be given. Both methods will be illustrated in the following examples.

Example 2.1. A mass m suspended by a spring of stiffness k. (Fig. 1.2)
We use the first method described above.

$$m\ddot{q}=mg-F=mg-kq.$$

At the position of equilibrium $q=mg/k$ so we put $q=mg/k+x$. Then

$$m\ddot{x}+kx=0.$$

This is already in the form (2) so that the mass oscillates with simple harmonic motion of period $2\pi\sqrt{m/k}$. An increase in mass or decrease in spring stiffness slows the oscillation.

Example 2.2. Combinations of springs. For the system of Fig. 2.2 the force on the mass due to the springs is the sum of

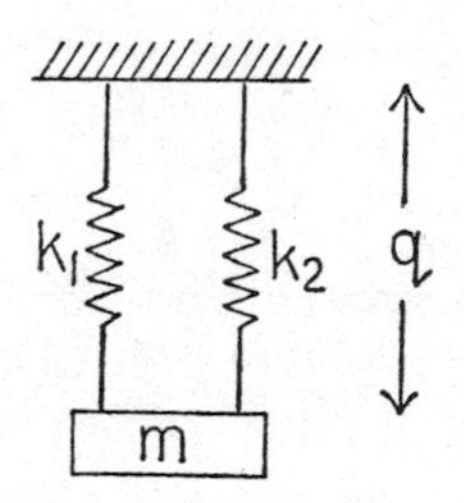

FIG. 2.2
Springs in parallel

the separate spring forces. Hence, if a_1 and a_2 are the natural lengths,

$$m\ddot{q} = mg - k_1(q - a_1) - k_2(q - a_2).$$

On substituting $q = (mg + k_1a_1 + k_2a_2)/(k_1 + k_2) + x$ we obtain

$$m\ddot{x} + (k_1 + k_2)x = 0.$$

The period of the simple harmonic motion is $2\pi\sqrt{\{m/(k_1 + k_2)\}}$. This result can be interpreted as: *the equivalent stiffness of two*

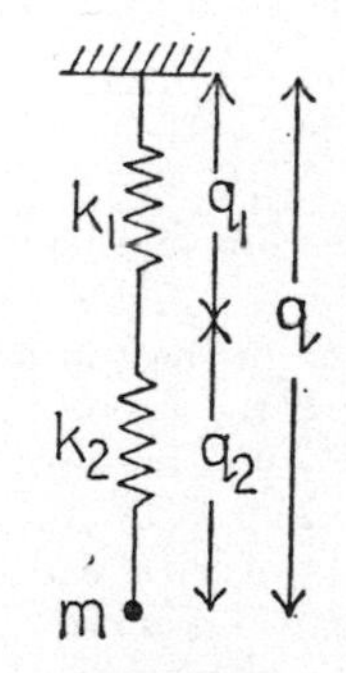

FIG. 2.3
Springs in series

springs in parallel is $k_1 + k_2$. The rule can obviously be extended to any number of springs in parallel.

On the other hand, for the system of Fig. 2.3, the tension is uniform throughout so that $k_1(q_1 - a_1) = k_2(q_2 - a_2)$ and

$$m\ddot{q} = mg - k_2(q_2 - a_2).$$

But $q = q_1 + q_2$ and so $(k_1 + k_2)q_2 = k_1(q - a_1) + k_2 a_2$. Therefore

$$m\ddot{q} = mg - \frac{k_1 k_2}{k_1 + k_2}(q - a_1 - a_2).$$

On putting $q = a_1 + a_2 + mg(k_1 + k_2)/k_1 k_2 + x$ we obtain

$$m\ddot{x} + \frac{k_1 k_2}{k_1 + k_2} x = 0.$$

This is the same equation as that of Ex. 2.1 with k replaced by $k_1 k_2/(k_1 + k_2)$ or $1/k$ replaced by $1/k_1 + 1/k_2$. Hence *the reciprocal of the equivalent stiffness of two springs in series is* $1/k_1 + 1/k_2$.

Example 2.3. The simple pendulum.

The simple pendulum (Fig. 2.4) consists of a light rigid rod (or inextensible string) smoothly pivoted at one end and carrying a particle of mass m at the other. Let l be the length of the rod

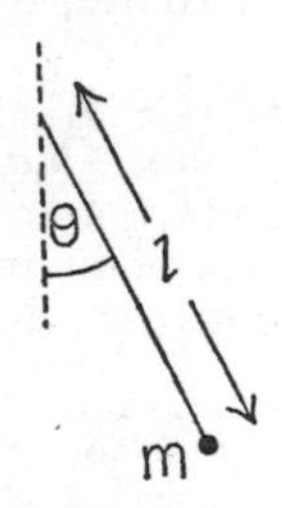

FIG. 2.4
The simple pendulum

and let θ be the inclination of the rod to the downward vertical.

In this example we use the second method detailed above. The kinetic energy $T = \frac{1}{2}ml^2\dot{\theta}^2$ and the potential energy $V = -mgl\cos\theta$. For small oscillations near the downward vertical expand in powers of θ, $\dot{\theta}$ and retain only terms up to the second order. No expansion is required in the kinetic energy because it is already in this form. For the potential energy we have

$$V = -mgl(1 - \tfrac{1}{2}\theta^2).$$

To this approximation the equation of the conservation of energy is

$$\tfrac{1}{2}ml^2\dot{\theta}^2 - mgl(1 - \tfrac{1}{2}\theta^2) = \text{constant}.$$

A time derivative gives, after cancellation of common factors,

$$l\ddot{\theta} + g\theta = 0.$$

The pendulum oscillates in a simple harmonic motion of period $2\pi\sqrt{l/g}$.

Example 2.4. Torsional oscillations.

When one end of a cylindrical shaft is clamped the shaft resists being twisted about its axis. A torsional couple or torque must be applied in order to turn it. It is found experimentally that, if the lower end is twisted through an angle ϕ, the applied

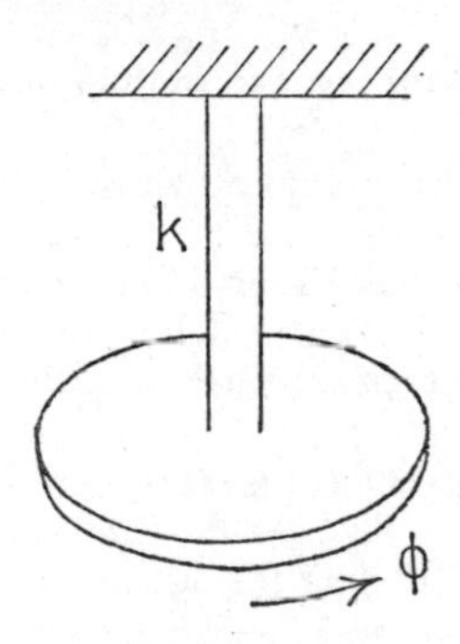

FIG. 2.5
Torsional oscillations

torque must be $k\phi$ provided that ϕ is small. The constant k is called the *torsional stiffness* of the shaft. Thus when the shaft is twisted a torque $k\phi$ is produced in it acting in such a way as to reduce ϕ.

Let a disc be firmly attached to the lower end of the shaft and let I be the moment of inertia of the disc about an axis perpendicular to its plane and along the axis of the shaft. Then

$$\begin{aligned} I\ddot{\phi} &= \text{torque increasing } \phi \\ &= -k\phi. \end{aligned}$$

Therefore, in small torsional oscillations, there is simple harmonic motion of period $2\pi\sqrt{I/k}$.

Example 2.5. In this example we consider the system of Fig. 2.6 where a spring of stiffness k is attached to a point of the

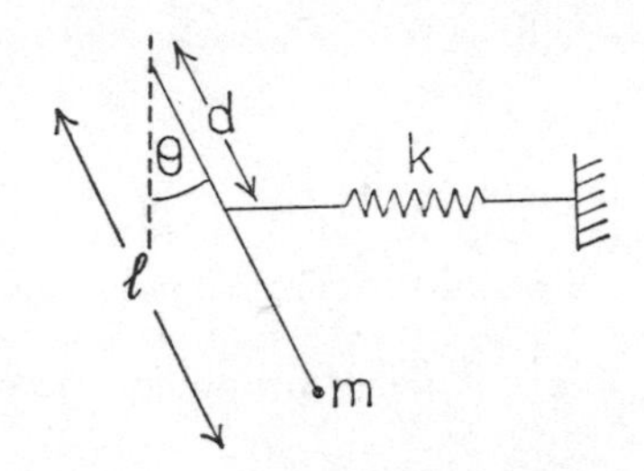

FIG. 2.6
The spring is equivalent to another one acting on *m*

light rigid rod of a simple pendulum. The other end of the spring is joined to a vertical wall in such a way that the spring always remains horizontal. The distance of the wall is chosen so that the spring has its natural length when $\theta = 0$. Then the vertical position of the pendulum is still a position of equilibrium.

The potential energy of the system

$$V = -mgl\cos\theta + \tfrac{1}{2}kd^2\sin^2\theta.$$

Expanding in powers of θ as far as the second order we obtain

$$V = -mgl(1 - \tfrac{1}{2}\theta^2) + \tfrac{1}{2}kd^2\theta^2.$$

Hence

$$\tfrac{1}{2}ml^2\dot{\theta}^2 - mgl(1 - \tfrac{1}{2}\theta^2) + \tfrac{1}{2}kd^2\theta^2 = \text{constant}$$

expresses the conservation of energy. Taking a time derivative we have

$$l\ddot{\theta} + (g + kd^2/ml)\theta = 0.$$

The system oscillates in simple harmonic motion of period $2\pi l\sqrt{\{m/(mgl + kd^2)\}}$. If we replace d by l and k by kd^2/l^2 this result is unaltered. Therefore the action of the spring is equivalent to that of a spring of stiffness kd^2/l^2 *attached to the mass*. Obviously, the spring has a very small effect if d is small.

Example 2.6. Transverse oscillations of a particle on a string.

A particle is attached to a point of a string stretched so tightly between two fixed points A and B that small changes in length cause no change in tension. With these high tensions the effect of gravity can be neglected but we include it for the sake of completeness.

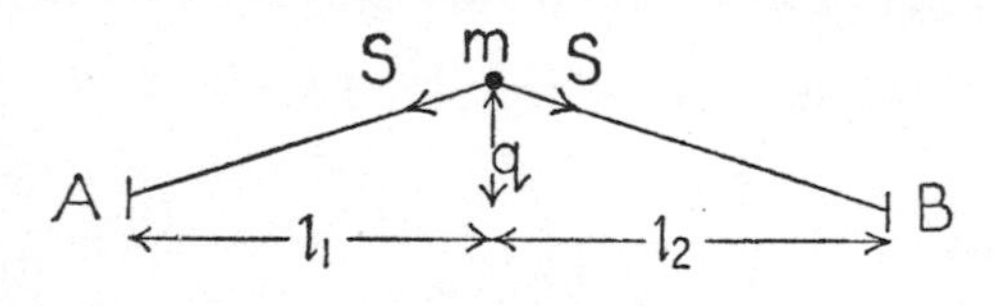

FIG. 2.7
Transverse oscillations

Let S be the constant tension in the string and q the distance of m from AB (we assume the motion of m is smoothly constrained to be perpendicular to AB). The force perpendicular to AB due to the left-hand portion of the string is $S \sin m\hat{A}B = S \tan m\hat{A}B$ correct to the first order if the angle $m\hat{A}B$ is small. Thus the force is Sq/l_1. Similarly the force duc to the right-hand portion is Sq/l_2. Hence

$$m\ddot{q} = -mg - Sq/l_1 - Sq/l_2.$$

Putting $q = -mgl_1l_2/S(l_1+l_2)+x$ we have

$$m\ddot{x} + S\left(\frac{1}{l_1}+\frac{1}{l_2}\right)x = 0$$

which is a simple harmonic motion of period $2\pi\sqrt{\{ml_1l_2/S(l_1+l_2\}}$.

The example serves as a crude model of what happens when a violin string is plucked. If we imagine the mass of the violin string concentrated at its mid-point then $l_1 = l_2 = \frac{1}{2}l$ and the period of oscillation is $\pi\sqrt{(ml/S)}$. According to this model the period of a violin string will vary as the square root of its length, as the square root of its mass and inversely as the square root of the tension. These variations were verified experimentally over 300 years ago by Mersenne. The magnitude of the period, however, is not correct which is scarcely surprising in view of the drastic replacement of mass distributed over the whole string by a particle.

Example 2.7. The electrical circuit.

When an inductance L and capacitance C are connected in series (Fig. 2.8) the equation governing the current I is, as is discussed more fully in Chapter IV,

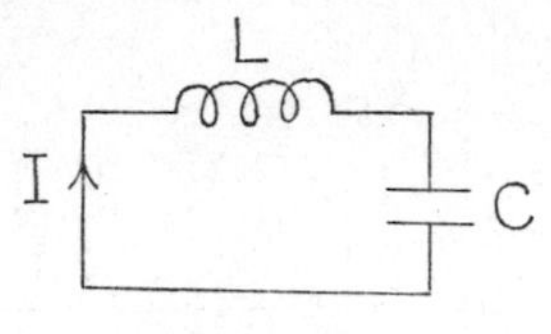

FIG. 2.8
Series circuit

$$L\frac{d^2I}{dt^2}+\frac{I}{C}=0.$$

The equation for the charge Q on the capacitor is

$$L\frac{d^2Q}{dt^2}+\frac{Q}{C}=0.$$

Both the charge and current vary in simple harmonic fashion with period $2\pi\sqrt{LC}$.

2.3. Forced oscillations. So far we have been considering the behaviour of a system when it is given a small initial disturbance and then left to itself. Now we turn our attention to the effect of a continuously applied disturbance. Let us consider in particular the system of Ex. 2.1. There is no loss of generality in limiting ourselves to this problem because all the other systems we have described satisfy the same differential equation. The only difference lies in our interpretation of the physical significance of the various quantities occurring in the equation.

If we imagine that someone is alternately pushing and pulling on the mass in Ex. 2.1 an extra term must be added to the equation of motion to account for the resulting force. Suppose that this force is $F_0 \cos(\omega t+\beta)$ (Fig. 2.9). The equation of motion becomes

$$m\ddot{x}+kx=F_0 \cos(\omega t+\beta) \tag{8}$$

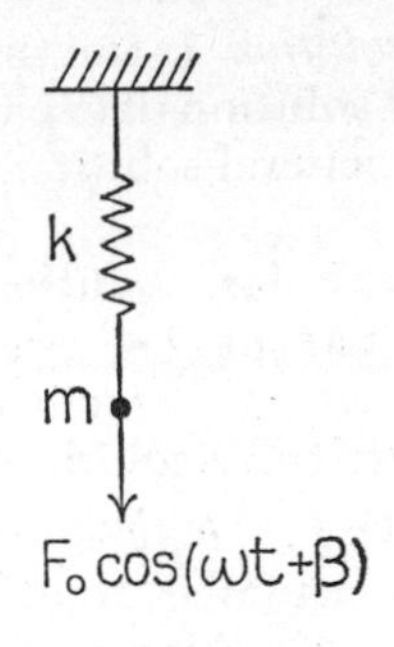

Fig. 2.9
The standard system

where (as before) x is the deflection from equilibrium. The same sort of equation arises in Exs. 2.2 and 2.6 if a harmonic vertical force is applied to the masses, in Exs. 2.3 and 2.5 if a harmonic horizontal force is applied, in Ex. 2.4 if a harmonic torque is applied and in Ex. 2.7 if a harmonic generator is connected in the circuit. It is therefore sufficient to discuss (8).

It might be thought to be a severe limitation to consider only continuous disturbances of the type appearing on the right-hand side of (8). There are two reasons why it is not. The first is that there are many perturbations of practical moment which are substantially harmonic. The second, and more important, reason is that there exists a theorem, known as *Fourier's theorem*,† which states that almost any disturbance can be represented as a sum or integral of harmonic terms. By adding together solutions of (8) with different ω's we can thereby solve the problem of a general applied force.

On dividing (8) by m and writing $k/m=\Omega^2$ we obtain

$$\ddot{x}+\Omega^2 x=(F_0/m)\cos(\omega t+\beta). \tag{9}$$

The method of finding the general solution of (9) is given in **RI 2.2** and **RI 2.5.** The general solution breaks into two

† See I. N. Sneddon's *Fourier Series* in this library.

parts—a *particular solution* (sometimes called particular integral), which is any solution of (9), and the *complementary function*, which is the general solution of (9) with zero right-hand side.

To derive a particular solution we observe that $\cos(\omega t+\beta)$ is the real part of $e^{i(\omega t+\beta)}$, i.e.

$$\cos(\omega t+\beta)=\mathscr{R}e^{i(\omega t+\beta)}.$$

We now look for a particular solution of

$$\ddot{\xi}+\Omega^2\xi=(F_0/m)\,e^{i(\omega t+\beta)}$$

and then say that a particular solution of (9) is $x=\mathscr{R}\xi$. Substitute $\xi=A\,e^{i(\omega t+\beta)}$. After cancelling the common factor $e^{i(\omega t+\beta)}$ we obtain

$$(\Omega^2-\omega^2)A=(F_0/m).$$

Hence our particular solution is

$$x=\mathscr{R}\frac{F_0/m}{\Omega^2-\omega^2}\,e^{i(\omega t+\beta)}=\frac{F_0/m}{\Omega^2-\omega^2}\cos(\omega t+\beta).$$

The complementary function has already been found because (9) is the equation of simple harmonic motion when the right-hand side is zero. Adding together the complementary function and particular solution we acquire, as the general solution of (9),

$$x=C\cos\Omega t+D\sin\Omega t+\frac{F_0/m}{\Omega^2-\omega^2}\cos(\omega t+\beta). \quad (10)$$

The solution fails if $\Omega=\omega$ because then $e^{i\omega t}$ occurs in the complementary function. In this case we must substitute $\xi=At\,e^{i(\omega t+\beta)}$ for a particular solution. Then $A=F_0/2im\omega$ so that $\mathscr{R}\xi=(F_0t/2m)\sin(\omega t+\beta)$ and the general solution of (9) is

$$x=C\cos\Omega t+D\sin\Omega t+\frac{F_0t}{2m\omega}\sin(\omega t+\beta).\ (\omega=\Omega) \quad (11)$$

The first two terms of (10) and (11) are always present whether there is a disturbing force or not. Their frequency $\Omega/2\pi$ is a characteristic of the system itself and their amplitude is determined only when the initial conditions are known. They constitute the *free oscillation.*

The last term in (10) and (11) arises only when there is an applied force. Its frequency $\omega/2\pi$ is the same as that of the applied force. The amplitude and phase do not depend in any way on the initial state of the system. This term is known as the *forced oscillation.*

If $\Omega^2 > \omega^2$ the phase of the forced oscillation has the same phase as the applied force, i.e. the mass is below equilibrium when the force pushes downwards. If $\Omega^2 < \omega^2$ write

$$\frac{F_0/m}{\Omega^2 - \omega^2} \cos(\omega t + \beta) = \frac{F_0/m}{\omega^2 - \Omega^2} \cos(\omega t + \beta + \pi);$$

there is now a phase difference of π between the forced oscillation and applied force, i.e. the mass is above equilibrium when the force pushes downwards.

The amplitude X_0 of the forced oscillation is $X_0 = F_0/m|\Omega^2 - \omega^2|$; a graph of the variations of X_0 with ω appears in Fig. 2.10. When $\omega \approx 0$ changes in the applied

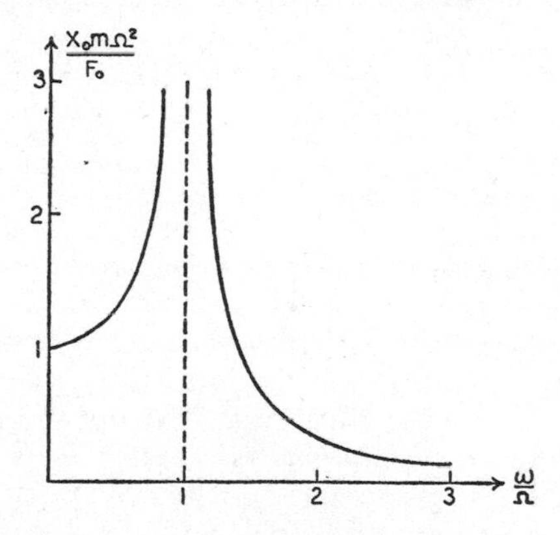

FIG. 2.10
Amplitude of forced oscillation

force take place very slowly. We would therefore expect the deflection of the mass at any instant to be the same as would occur under a constant force equal to the instantaneous value of the applied force. The displacement of the mass under the constant force F_0 would be $F_0/k = F_0/m\Omega^2$. Thus near $\omega = 0$ we expect $X_0 m\Omega^2/F_0$ to be near 1 and Fig. 2.10 shows that, indeed, it is. On the other hand if $\omega/\Omega >> 1$ the applied force will be changing direction so rapidly that the mass will have no opportunity to follow it. The amplitude of the forced oscillation should then be small and this is borne out by Fig. 2.10.

When $\omega = \Omega$ the direction of the applied force changes at the same rate as the direction of m in its free oscillation. Therefore as the mass is moving downwards the force will be pushing downwards and both will be in the upwards direction at the same time. An indefinitely large amplitude of oscillation should then build up. In fact Fig. 2.10 displays an infinity at $\omega = \Omega$ but this is because (10) has been employed in drawing the graph and we have already explained that (10) must be replaced by (11) when $\omega = \Omega$. Nevertheless the presence of the factor t in (11) shows that the amplitude of the forced oscillation increases steadily to an infinite value. The phenomenon is known as *resonance*. Sometimes the natural frequency $\Omega/2\pi$ is known as the resonant frequency.

Of course, infinite amplitudes do not occur in nature. They are limited by some means or other which eventually modify the governing differential equation—indeed, for general systems, the differential equation has been derived only on the assumption of small displacement from equilibrium. When $\omega = \Omega$ this assumption will be violated after a certain time. In spite of this, large (but not infinite) amplitudes do occur in physical systems when ω is near Ω, as we shall see later on, which is why the phenomenon of resonance is so important.

When ω and Ω have general values the solution (10) consists of the sum of two sinusoidal terms of different frequencies. They can never be combined into a single sinusoidal term. Therefore the motion is not simple harmonic.

The diagram in Fig. 2.10 has been drawn on the assumption that F_0 itself does not involve ω. It may happen that F_0 does depend on ω. For example, in certain electrical circuits $F_0=\omega G_0$ where G_0 is independent of ω; the main difference to Fig. 2.10 is that the curve tends to zero as $\omega \to 0$. In examples concerning rotating machinery we may have $F_0=\omega^2 G_0$; then the amplitude of the forced oscillation tends to zero as $\omega \to 0$ and tends to the constant value G_0/m as $\omega \to \infty$. The phenomenon of resonance is still present.

The constants C and D in (10) are determined as soon as the initial conditions are known. Suppose that $x=x_0$ and $\dot{x}=x_1$ when $t=0$. Substitution in (10) gives

$$x=(x_0-X_0 \cos \beta) \cos \Omega t + \frac{x_1+X_0\omega \sin \beta}{\Omega} \sin \Omega t + X_0 \cos (\omega t+\beta) \tag{12}$$

where $X_0=F_0/m(\Omega^2-\omega^2)$. If ω is near Ω but not actually equal to it X_0 is large and the three terms involving X_0 in (12) dominate the rest. Retaining only the terms with X_0 we can write the solution as

$$X_0 \cos (\omega t+\beta) - X_0 \cos (\Omega t+\beta) + \frac{X_0(\omega-\Omega)}{\Omega} \sin \beta \sin \Omega t.$$

The factor $\omega-\Omega$ in the numerator of the last term makes it of no consequence in comparison with the first two. Hence the approximate solution is $2X_0 \sin\{\frac{1}{2}(\omega+\Omega)t+\beta\} \sin \frac{1}{2}(\Omega-\omega)t$. The first factor represents harmonic motion with a frequency of nearly $\Omega/2\pi$. The second factor varies very slowly since $\omega \approx \Omega$. Hence the solution can be interpreted as a harmonic motion of period $2\pi/\Omega$ whose amplitude slowly waxes and wanes. This is the phenomenon of ***beats***.

The solution (12) has been found by deriving the general solution of (9) and then imposing the initial conditions. By using p-operators one can obtain (12) directly from (9). We now illustrate this method. First recall the rules for p-operators (**RII 2.1** and **2.2**): (i) replace d/dt by p and add to the right-hand side the terms involving positive powers of p in

$$(p^2+\Omega^2)\left(x_0+\frac{x_1}{p}\right);$$

(ii) the operational representation of $t^m e^{\lambda t}/m!$ is

$$p/(p-\lambda)^{m+1}.$$

From (ii) it follows that the operational representations of $\cos \omega t$ and $\sin \omega t$ are $p^2/(p^2+\omega^2)$ and $p/(p^2+\omega^2)$ respectively. Hence the representation of $\cos(\omega t+\beta)$ is $(p^2 \cos\beta - p\omega \sin\beta)/(p^2+\omega^2)$. Therefore, remembering (1), we obtain for our operational equation

$$(p^2+\Omega^2)x=(F_0/m)\frac{(p^2\cos\beta-p\omega\sin\beta)}{p^2+\omega^2}+p^2x_0+px_1.$$

Hence

$$x=\frac{(F_0/m)(p^2\cos\beta-p\omega\sin\beta)}{(p^2+\omega^2)(p^2+\Omega^2)}+\frac{p^2x_0+px_1}{p^2+\Omega^2}$$

$$=X_0(p^2\cos\beta-p\omega\sin\beta)\left(\frac{1}{p^2+\omega^2}-\frac{1}{p^2+\Omega^2}\right)+\frac{p^2x_0+px_1}{p^2+\Omega^2}$$

and we recover (12) at once.

EXERCISES ON CHAPTER II

1. Find the periods of small oscillations about the symmetrical positions of stable equilibrium in Exercises 3, 6 and 2 of Chapter I.

2. A light rigid rod of length l is smoothly pivoted at one end and carries a bob of mass m. A horizontal spring of stiffness k is attached to the bob, the other end being attached to a fixed wall. If the spring is unstrained when the rod is vertical find the period of small oscillations.

3. A light rigid rod of length $4l$ is smoothly pivoted at its centre. A particle of mass $2m$ is carried on its upper end and one of mass $3m$ on its lower end. A horizontal spring of stiffness k is connected to the rod at a distance $\frac{1}{2}l$ below the pivot, the other end being connected to a fixed wall. If the spring is unstrained when the rod is vertical find the period of small oscillations.

4. A light rigid rod is smoothly pivoted at one end. At a distance l from the pivot is a particle of mass m, and, at a distance $3l$, there is a spring of stiffness k perpendicular to the rod. The system is in equilibrium with the rod horizontal. Show that the period of small oscillations is $\frac{2}{3}\pi\sqrt{(m/k)}$.

5. Two equal horizontal circular cylinders, with their axes a distance $2d$ apart, rotate with great angular speed in opposite directions; the one of the left rotates in a clockwise sense and that on the right anti-clockwise. A horizontal rod of mass m is at rest on top of the cylinders with its centre of mass midway between the axes. The rod is displaced a small distance a and then released. Show that its subsequent displacement is $a \cos t \sqrt{(\mu g/d)}$ where μ is the coefficient of Coulomb dry friction (i.e. the frictional force of a cylinder is μR where R is the pressure on it).

6. The combined mass of a grindstone and its axle is M and the moment of inertia about the axis of rotation is I. The grindstone is slightly unbalanced, i.e. the centre of mass does not lie on the axis of rotation. The grindstone is lowered through a slot in a horizontal table so that the system is at rest with only the axle in contact with the table. When the system is slightly disturbed it rolls in simple harmonic motion of slow frequency $\Omega/2\pi$. Show that the stone can be balanced by the addition of a mass m at a distance of $(I+Mr^2)\Omega^2/mg$ approximately from the axis of rotation, r being the radius of the axle.

7. A solid uniform circular disc of radius 40 inches can rotate freely about a horizontal axis, perpendicular to its plane, passing through its centre C. A circular hole, with its centre distant 12 inches from C, of radius 8 inches is cut in the disc. Find the period of small oscillations.

8. A disc, of moment of inertia 800 inch-poundals. sec.^{2} about its axis, is firmly connected to a solid steel shaft of diameter $1\frac{1}{2}''$ and length $25''$. At the other end of the shaft is a gear of diameter $5''$, which meshes with a gear, of diameter $2\frac{1}{2}''$, on a solid steel shaft of diameter $2''$ and length $25''$. The second shaft carries a disc of moment of inertia 320 inch-poundals. sec.^{2} at its end. In torsional motion the angles turned through by the discs are ϕ_1 and ϕ_2 respectively. Show that the variation of $\phi_2 - 2\phi_1$ is simple harmonic with period 0·052 secs. Assume that the gears do not slip and have no inertia. [The

torsional stiffness of solid steel shaft is $37{\cdot}8 \times 10^6\, d^4/l$ inch-poundals where the diameter d and length l are measured in inches.]

9. A light elastic string passing over a smooth peg has masses M and m attached to its ends. The system is released from rest with the string just slack. Prove that, after a time t, the tension (assumed to be uniform throughout the string) is $2mMg(1 - \cos \Omega t)/(m+M)$ where $\Omega^2 = \lambda(M+m)/aMm$, λ and a being the modulus of elasticity and natural length of the string respectively.

10. A particle is supported by a vertical spring below it, the other end being attached to a weightless plane P. The spring is compressed 3 inches under the weight of the particle. P stands on a platform which is initially at rest and then moves downwards with acceleration $4g$. How far does the platform go before P leaves it?

11. A uniform rod of mass 1 lb. and length 3 ft., is smoothly pivoted at one end. A horizontal force, of amplitude 4 poundals and varying sinusoidally with frequency $\frac{1}{2}$ cycle per second, is applied at a distance of 9″ from the pivot. Find the amplitude of the angular forced oscillation.

12. A particle of mass m is at rest at the end of a spring of stiffness k hanging from a fixed platform. At $t=0$ a constant downwards force F is applied to the mass. The force is removed at $t=t_0$. Prove that thereafter the displacement of the mass from equilibrium is

$$(F/k)\{\cos \Omega(t-t_0) - \cos \Omega t\}$$

where $\Omega^2 = k/m$.

13. A particle of mass m is suspended by a spring of stiffness k from a platform. The platform is obliged to oscillate in simple harmonic motion of amplitude a and frequency $\omega/2\pi$. Find the amplitude of the forced oscillation of the mass *relative to the platform*.

The amplitude of the forced oscillation is found to be 4″ when $\omega = 10\pi$. What limitation is there on k in order that $a = 4''$ to within 10 per cent, it being known that $k/m < 100\pi^2$.

14. A circular disc, of mass 10 lb. and radius 10″, is mounted on a solid steel shaft of diameter $\frac{1}{2}''$ and length 30″. The disc is subject to a simple harmonic torque of magnitude M ft.-poundals and frequency 8 cycles per second. Find the amplitude of the forced oscillation. (For torsional stiffness see Q. 8.)

CHAPTER THREE

The Effect of Friction

3.1. Introduction. Up to this stage attention has been centred on the behaviour of a system near equilibrium in the absence of dissipative forces. Though these forces may be small they are always present in any natural system and cause a gradual diminution of the oscillations. Friction can take many forms as the reader will see from the following brief description of various aspects.

The friction between two moving dry bodies in contact was investigated over two centuries ago by Coulomb. The results of his researches are mbodied in the law $F=\mu R$, familiar in school books, that the friction is proportional to the normal pressure. There is no mention of the velocities of the bodies in this law but it has been found that the friction depends on the relative velocities of the bodies. In fact the friction decreases from its Coulomb value as the relative velocity of the bodies increases. When this variation is taken into account we have the law of *dry friction*; graphical presentation is usually necessary.

If oil or other lubricating fluid is introduced between the surfaces the situation changes radically. In the ideal case of *perfect lubrication* the friction is proportional to the relative velocity of the bodies but is independent of the normal pressure. For most bodies in contact the frictional effects will be somewhere between the extremes of dry friction and perfect lubrication.

When a body vibrates in a fluid such as air, the *fluid resistance* causes a diminution of the motion. For small relative velocities the force is proportional to the relative velocity and nearly proportional to the square at higher relative velocities.

Even the absence of fluid resistance does not imply the non-existence of damping because of the possibility of *internal friction.* This will cause damping of an elastic body even in a vacuum. The force is sometimes nearly proportional to the velocity.

Sufficient has been said to indicate the complexity of frictional phenomena and to demonstrate that there is little likelihood of dealing with all types of friction by one simple general law. In order to obtain any clue as to the effect of friction on oscillations it is therefore necessary to choose the simplest law consistent with a number of practical applications. Accordingly we limit ourselves to friction which is proportional to the relative velocity.

It is convenient to have available a standard device which will produce friction proportional to the relative velocity. This device, which is displayed pictorially in Fig. 3.1, is known as a *dashpot.* It can be imagined as a plunger moving in a pot of oil. If the pot is kept fixed and

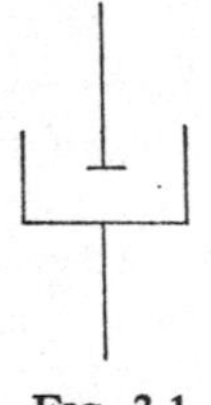

FIG. 3.1
The dashpot

the plunger is pushed in at speed u the motion of the plunger is opposed by a force Du where D is a positive constant called the *coefficient of damping*; an equal and opposite force is, of course, transmitted to the pot. On the other hand, if the plunger is kept fixed and the pot is pulled down with speed u the motion of the pot is opposed by a force Du. More generally, if the plunger moves down with speed u and the pot with speed v the motion of the plunger is opposed by a force $D(u-v)$ and that of the

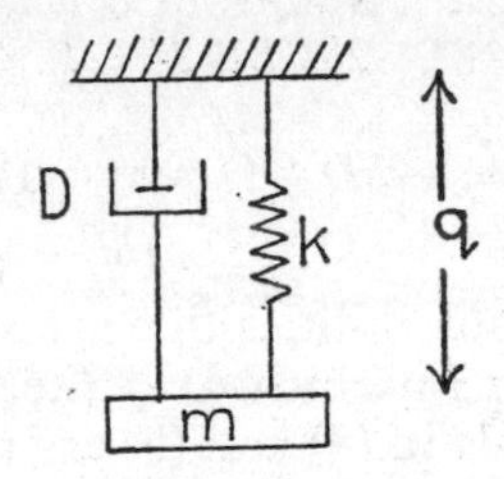

FIG. 3.2
The standard system with viscous damping

pot by a force $D(v-u)$. Damping of the type due to a dashpot is frequently known as *viscous damping.*

3.2. Free oscillations with viscous damping. To consider the effect of viscous damping we take our standard system of a mass m and a spring of stiffness k and add a dashpot between the mass and fixed platform. Let q be the depth of the mass below the platform. Then, at any instant, the plunger of the dashpot is fixed and the pot moving down with speed $\dot{q}$. Consequently, the motion of the pot is opposed by a force $D\dot{q}$. The equation of motion of the system is

$$m\ddot{q}=mg-k(q-a)-D\dot{q}$$

where a is the natural length of the spring. Substituting $q=a+x+mg/k$ where x is the displacement from equilibrium we obtain

$$m\ddot{x}+D\dot{x}+kx=0.$$

Divide through by m and write $k/m=\Omega^2$, $D/m=2(D/2\sqrt{km})\sqrt{(k/m)}=2b\Omega$ where

$$b=D/2\sqrt{km}. \qquad (1)$$

The equation becomes

$$\ddot{x}+2b\Omega\dot{x}+\Omega^2x=0. \qquad (2)$$

As in **RI 2.1** the general solution of (2) is

$$x=A\,e^{\delta_1 t}+B\,e^{\delta_2 t} \qquad (3)$$

where δ_1 and δ_2 are the values of δ which satisfy

$$\delta^2+2b\Omega\delta+\Omega^2=0.$$

Thus

$$\delta_1=-b\Omega+\Omega\sqrt{(b^2-1)}, \tag{4}$$

$$\delta_2=-b\Omega-\Omega\sqrt{(b^2-1)}. \tag{5}$$

There are three possibilities to consider:
(*a*) Strong damping in which $b>1$ (i.e. $D^2>4km$). In this case both the radicals in (4) and (5) are real and less than b. Hence both δ_1 and δ_2 are real and *negative*; also $\delta_1>\delta_2$. Both exponentials in (3) decrease steadily as t increases. Whether x grows or falls initially depends upon the signs and magnitudes of A and B but it is not difficult to verify that x has at most one stationary value and that there is at most one value of t at which $x=0$. Typical possibilities are shown in Fig. 3.3.

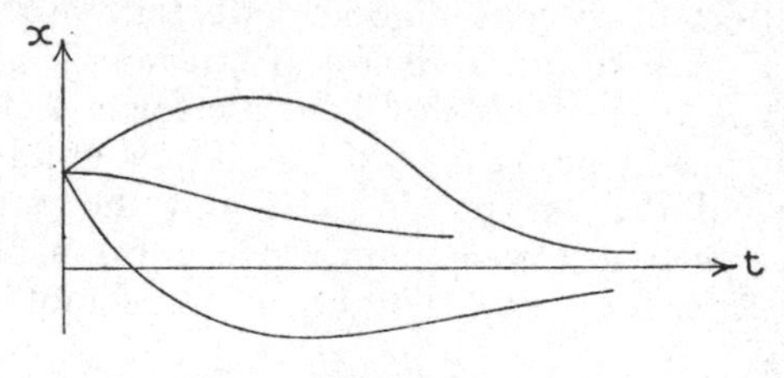

FIG. 3.3
Strong damping

The effect of damping here has been to destroy completely the oscillations; the motion is more of a crawling towards equilibrium. This property is very useful in the design of measuring instruments which are required to make only one swing to their equilibrium position.

There is one particular case of strong damping to be commented on and that arises when $m=0$. Then the derivation of (2), which involves division by m, is no longer valid and we must return to our original equation which takes the form

$$D\dot{x}+kx=0.$$

Therefore $x=x_0\, e^{-kt/D}$ which shows that the magnitude of x decreases steadily. In time D/k the displacement reduces to $1/e$th of its initial value. For this reason D/k is known as the *time constant* or *relaxation* of the system.

(b) Critical damping in which $b=1$ (i.e. $D^2=4km$). Here the solution (3) fails because $\delta_1=\delta_2$. The quadratic now has the double root $-\Omega$ and so

$$x=(A+Bt)e^{-\Omega t}.$$

The motion is not greatly different from that shown in Fig. 3.3. There is no oscillation and the system passes through equilibrium at most once.

(c) Light damping in which $b<1$ (i.e. $D^2<4mk$). The radicals in (4) and (5) are now purely imaginary. Write $\Omega\sqrt{(b^2-1)}=in$ where $n=\Omega\sqrt{(1-b^2)}$; n is real. Then

$$\begin{aligned} x&=e^{-b\Omega t}(A'\cos nt+B'\sin nt)\\ &=X\,e^{-b\Omega t}\cos(nt+\epsilon). \end{aligned}$$

The solution consists of two factors, a decreasing exponential and a sinusoidal term. The combined result is a sine wave of decreasing amplitude lying between the exponen-

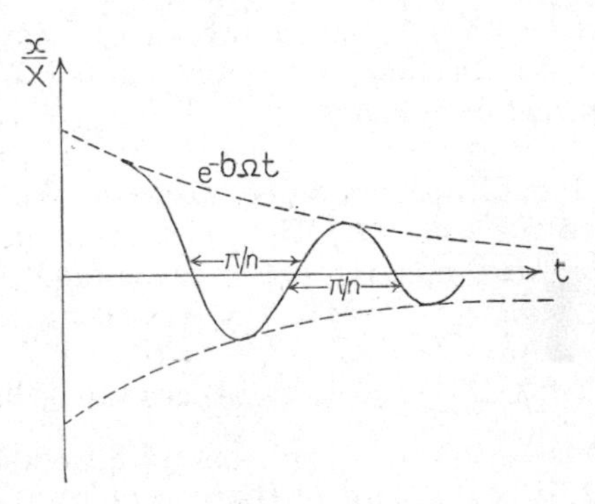

FIG. 3.4
Damped oscillation

tial curve and its mirror image (Fig. 3.4). The system passes through the equilibrium position $x=0$ an infinite number of times. consecutive occasions being separated by a time interval of π/n. For this reason the motion is called a *damped oscillation* of period $2\pi/n$.

The damped oscillation has a maximum x_q when $nt+\epsilon=2q\pi$, q being an integer. The time interval between consecutive maxima is $2\pi/n$.

Now

$$\frac{x_q}{x_{q+1}}=\frac{e^{-(b\Omega/n)(2q\pi-\epsilon)}}{e^{-(b\Omega/n)\{2(q+1)\pi-\epsilon\}}}=e^{2b\Omega\pi/n}$$

which is independent of q, t or X, i.e. the ratio of two consecutive maxima is a constant which is characteristic of the system. The quantity $2b\Omega\pi/n=D\pi/mn$, which is a measure of the rate of decrease of amplitude, is called the *logarithmic decrement.*

If the damping is very small $b<<1$ and then $n=\Omega$ with an error of the second order. Thus the introduction of slight damping does not alter the natural frequency of a system performing free oscillations but does affect the amplitude by bringing in a slowly decreasing exponential factor.

The value of $D(=2\sqrt{mk})$ at which the transition from oscillatory to non-oscillatory motion occurs is known as the *critical coefficient of damping.*

3.3. Forced oscillations with viscous damping. If a vertical force $F_0 \cos(\omega t+\beta)$ be applied to the particle of Fig. 3.2 the equation of motion must be modified to

$$m\ddot{x}+D\dot{x}+kx=F_0\cos(\omega t+\beta)$$

or

$$\ddot{x}+2b\Omega\dot{x}+\Omega^2x=(F_0/m)\cos(\omega t+\beta). \qquad (6)$$

The general solution of this differential equation (**RI 2.2, I 2.5**) is the sum of the complementary function and a particular solution. The complementary function, which is the solution of (6) with zero right-hand side, is the free damped motion of the preceding section. For a particular solution replace $\cos(\omega t+\beta)$ by $e^{i(\omega t+\beta)}$ and look for a solution of

$$\ddot{\xi}+2b\Omega\,\dot{\xi}+\Omega^2\xi=(F_0/m)\,e^{i(\omega t+\beta)}.$$

Try $\xi=A\,e^{i(\omega t+\beta)}$; then

$$(-\omega^2+2ib\Omega\omega+\Omega^2)A=F_0/m.$$

Hence a particular solution of (6) is $\mathscr{R}\xi$

$$=\mathscr{R}\frac{(F_0/m)\,e^{i(\omega t+\beta)}}{\Omega^2-\omega^2+2ib\Omega\omega}$$

$$=\frac{F_0/m\,\{(\Omega^2-\omega^2)\cos(\omega t+\beta)+2b\Omega\omega\sin(\omega t+\beta)\}}{(\Omega^2-\omega^2)^2+4b^2\Omega^2\omega^2}.$$

Consequently

$$x=\text{free damped oscillation}+\frac{(F_0/m)\cos(\omega t+\beta-\phi)}{\{(\Omega^2-\omega^2)^2+4b^2\Omega^2\omega^2\}^{1/2}} \qquad (7)$$

where

$$\cos\phi=\frac{\Omega^2-\omega^2}{\{(\Omega^2-\omega^2)^2+4b^2\Omega^2\omega^2\}^{1/2}},$$

$$\sin\phi=\frac{2b\Omega\omega}{\{(\Omega^2-\omega^2)^2+4b^2\Omega^2\omega^2\}^{1/2}}. \qquad (8)$$

The first term of (7) is always present whether there is an applied force or not. Its magnitude depends upon the initial conditions but its amplitude, whatever its initial value, diminishes to zero as t becomes large enough. For this reason the first term of (7) is frequently known as a *transient*.

The second term of (7), which is simple harmonic, only arises when there is an applied force. Its amplitude is independent of the initial conditions and its frequency is the same as that of the applied force. It is known as the *forced oscillation.*

Often it is only what happens after some length of time that is of interest. Then the transient can be discarded and only the forced oscillation considered. But this is not always true—the behaviour for all times may be important. The transient cannot then be neglected (it is present even if $x=0$, $\dot{x}=0$ when $t=0$). The effect of the transient is to prevent the displacement having the same simple harmonic variation as the applied force. This distortion of the waveshape can be troublesome particularly in servomechanisms.

The amplitude X_0 of the forced oscillation can be written, in dimensionless form,

$$X_0 = \frac{F_0/k}{\left\{\left(1 - \frac{\omega^2}{\Omega^2}\right)^2 + \frac{4b^2\omega^2}{\Omega^2}\right\}^{1/2}} \qquad (9)$$

which is plotted graphically for various values of b in Fig. 3.5. (It is worth remarking that the quickest way of finding X_0 in any given problem is to take the modulus of the complex quantity A above, i.e. the ratio of the moduli of the numerator and denominator of A.) Varia-

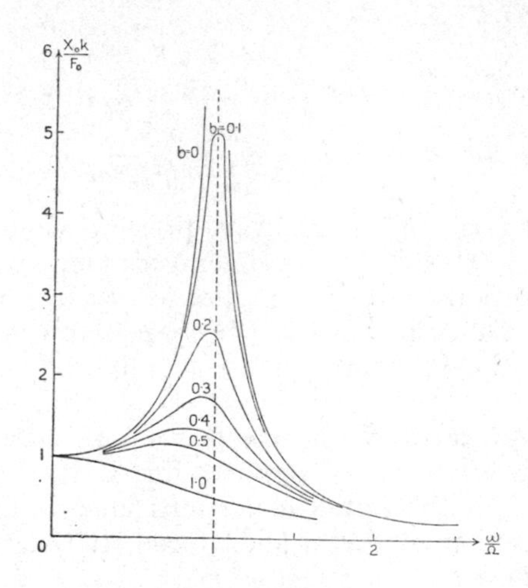

Fig. 3.5
Amplitude of forced oscillation for various dampings

tions of X_0 with ω/Ω occur only in the denominator and are dictated by the changes in

$$(1 - z^2)^2 + 4b^2z^2$$

if we put $z^2 = \omega^2/\Omega^2$. The derivative of this is

$$-4(1 - z^2)z + 8b^2z$$

which vanishes when $z = 0$ or $z^2 = 1 - 2b^2$. The second

derivative has the value $4(2b^2 - 1)$ when $z=0$ and the value $8(1 - 2b^2)$ when $z^2=1 - 2b^2$. Since z^2 cannot be negative, the denominator of X_0 has a minimum at $z=0$ and no other stationary point if $2b^2>1$, but a maximum at $z=0$ and a minimum at $z^2=1-2b^2$ if $2b^2<1$. Hence

if $2b^2>1$, X_0 has a maximum at $\omega=0$;

if $2b^2<1$, X_0 has a minimum at $\omega=0$ and a maximum at $\omega^2/\Omega^2=1-2b^2$.

The value of X_0 at $\omega^2/\Omega^2=1 - 2b^2$ is $\dfrac{F_0/k}{2b(1 - b^2)^{1/2}}$ which can also be written as F_0/Dn. This maximum increases as the damping decreases, becoming infinite when $b=0$.

The reader should note that there are three important frequencies now associated with the system:

(i) $\Omega/2\pi$ the undamped natural frequency.

(ii) $n/2\pi=(\Omega/2\pi)\sqrt{(1 - b^2)}$ the damped free frequency.

(iii) $(\Omega/2\pi)\sqrt{(1 - 2b^2)}$ the frequency at which the amplitude of the forced oscillation is a maximum and usually known as the *resonant frequency*. All these frequencies coincide when $b=0$ and differ only in the second order when b is non-zero but small, e.g. if $b<0{\cdot}1$ they all agree to within 1 per cent.

The phase difference ϕ between the forced oscillation and the applied force is plotted in Fig. 3.6. We observe, from (8), that $\cos\phi=0$, $\sin\phi=1$ and $\phi=\frac{1}{2}\pi$ when $\omega=\Omega$. Also, as $\omega/\Omega\to 0$, $\cos\phi\to 1$ and $\sin\phi\to 0$ so that $\phi\approx 0$. As $\omega/\Omega\to\infty$, $\cos\phi\to -1$ and $\sin\phi\to 0$ so that $\phi\to\pi$.

Certain salient points of the graphs in Figs. 3.5 and 3.6 can be obtained by physical reasoning. For slow vibrations the forces due to inertia and damping are small so that the applied force is used mainly in overcoming the spring force. Therefore the displacement is in phase with the applied force and equals the static extension under the instantaneous applied force. This is in agreement with the graphs near $\omega=0$. In fast vibrations the inertial force is very much larger than the others so that the applied force is used primarily in overcoming it. There will be a phase

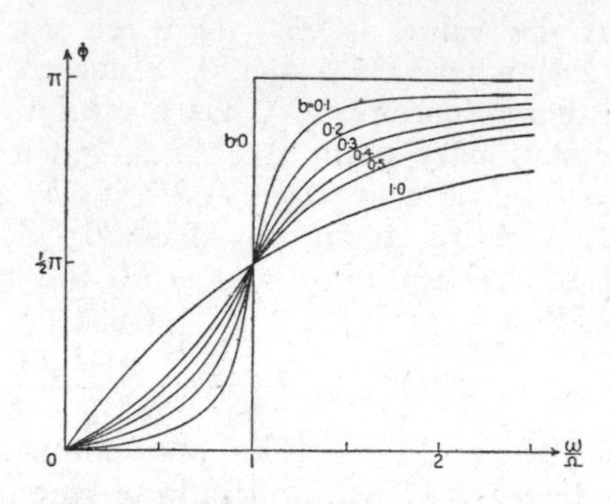

FIG. 3.6
The phase difference as a function of ω

change of π and the amplitude will be small, as is borne out by the graphs. At resonance the inertial and spring forces balance so that the applied force has only to overcome the damping. There is a phase change of $\frac{1}{2}\pi$ and an amplitude of $F_0/\omega D$.

3.4. The production of heat. In a damped system the applied force does work in producing an oscillation against the action of friction. This work is transformed into heat and it is important to know the amount of heat being produced in order that arrangements can be made to dissipate it and prevent the system becoming too hot. The heat produced by the transient motion can be neglected since it vanishes eventually.

The rate at which work is done by the applied force is

$$\dot{x}F_0 \cos(\omega t+\beta)$$
$$= -\omega X_0 F_0 \cos(\omega t+\beta) \sin(\omega t+\beta-\phi)$$

the transient being ignored. However, we are not interested so much in the instantaneous rate as the quantity produced over a period of time. To that end we calculate the average over the time $2\pi/\omega$ of a complete oscillation of the applied force. We obtain

$$-\frac{\omega}{2\pi}\int_0^{2\pi/\omega} \omega X_0 F_0 \cos(\omega t+\beta)\sin(\omega t+\beta-\phi)dt$$
$$=\frac{\omega^2 X_0 F_0}{4\pi}\int_0^{2\pi/\omega}\{\sin\phi-\sin(2\omega t+2\beta-\phi)\}dt$$
$$=\tfrac{1}{2}\omega X_0 F_0 \sin\phi.$$

On substituting from (8) and (9) we find for the average rate

$$\frac{F_0^2 b\Omega\omega^2/m}{(\Omega^2-\omega^2)^2+4b^2\Omega^2\omega^2}=\frac{\frac{1}{2}DF_0^2\omega^2}{(k-m\omega^2)^2+\omega^2D^2}. \qquad (10)$$

A graph of this quantity is shown in Fig. 3.7. There is a maximum of $\frac{1}{2}F_0^2/D$ at $\omega=\Omega$.

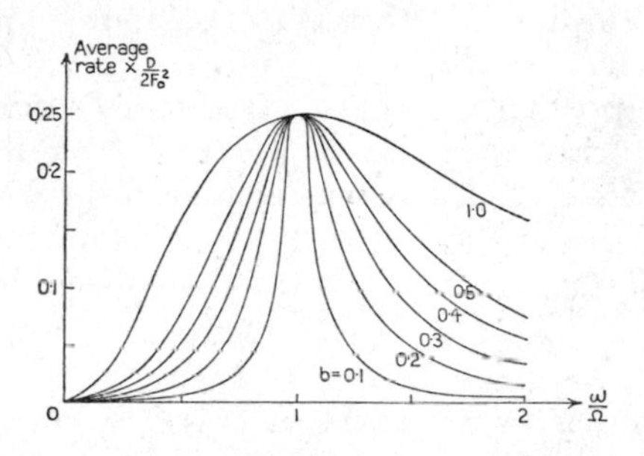

FIG. 3.7
Heat production

Observe that if $D=0$ there is no work done on the average because there is no friction to dissipate energy.

In c.g.s. units the quantity (10) is measured in ergs./ sec. The number of calories of heat developed in t seconds is obtained by multiplying (10) by $t\times10^{-7}/4{\cdot}18$. (In the f.p.s. system multiplication by $4t\times10^{-5}$ gives the heat in B.Th.U.)

3.5. Frequency measurement. The phenomenon of resonance forms the basis of the design of most frequency measuring instruments. The problem in frequency measurement is to determine the unknown frequency $\omega/2\pi$

of an applied sinusoidal force. If we adjust the natural frequency $\Omega/2\pi$ of our system until resonance occurs, i.e. the amplitude of the forced oscillations is a maximum we know from section 3.3 that Ω can be taken equal to ω provided that the damping is sufficiently small. The determination of Ω may be carried out either by calculation or by reference to a standard frequency source.

How Ω is altered in practice varies with the instrument. One device for mechanical oscillations consists of a steel strip with a small mass at the end. Adjustment of the natural frequency is achieved by changing the length of the strip. For electrical measurements the series circuit of Fig. 2.8 may be employed, either the capacitance or inductance being variable.

A similar problem to that of frequency measurement is picking one frequency out of several. Providing that the frequencies are sufficiently widely spaced the forced oscillation due to the one equal to the resonant frequency will dominate all the others. It is on this principle that radio and television receivers are designed.

3.6. Forces due to mountings. It is frequently important to mount oscillatory or rotating machines so that there is little or no vibration in the foundation, e.g. the motor-car engine and electric razor. Rubber mountings are frequently used and we can visualize these as forming a combination of spring and dashpot. A model for our problem is therefore provided by Fig. 3.2. We can imagine this as an oscillatory mass mounted on the ceiling or, on turning the diagram upside down, on the floor.

The force on the platform due to $F_0 \cos(\omega t+\beta)$ on m is the spring force kx and the damping force $D\dot{x}$ due to the plunger. Hence the total force on the platform is $D\dot{x}+kx=X_0\{k\cos(\omega t+\beta-\phi)-D\omega\sin(\omega t+\beta-\phi)\}$ the transient being neglected. The amplitude of this force is $X_0\sqrt{(k^2+D^2\omega^2)}=X_0k\sqrt{(1+4b^2\omega^2/\Omega^2)}$.

The efficiency of the mounting can be estimated by

dividing the square of this amplitude into F_0^2 the square of the amplitude of the applied force. This gives for the efficiency

$$\frac{F_0^2}{X_0^2k^2(1+4b^2\omega^2/\Omega^2)}$$
$$=\frac{(\Omega^2-\omega^2)^2+4b^2\omega^2\Omega^2}{\Omega^4+4b^2\omega^2\Omega^2}. \quad (11)$$

The larger this quantity the more efficient the mounting. The quantity (11) can be written as

$$1+\frac{\omega^2(\omega^2-2\Omega^2)}{\Omega^2(\Omega^2+4b^2\omega^2)}.$$

From this we see that if $\omega^2<2\Omega^2$ the efficiency is less than 1, i.e. the force transmitted to the platform is *greater* than the applied force. The situation improves as b increases but the ratio can never be greater than 1. In other words, if $\omega^2<2\Omega^2$, the effect of the spring mounting is so bad that damping cannot counteract it. On the other hand, if $\omega^2>2\Omega^2$, the efficiency is always greater than 1 but is reduced as b is increased from zero. Theoretically, the best efficiency can be achieved by using no damping ($b=0$) and a very soft spring ($\Omega << \omega$). However, the same efficiency can be obtained with $b\neq 0$ by choosing a somewhat softer spring. The advantage of such a procedure is that it ensures the damping of the free oscillation and lessens the risk if a sudden alteration of ω causes operation near resonance.

3.7. Amplitude measurement. In order to measure the amplitude of an oscillation such as an earthquake it is necessary to modify the preceding approach. A typical *seismograph* consists of a box containing an oscillatory mechanism; the motion of the mechanism is observed on a scale *attached to the box.*

Our model for the seismograph is a mass supported by a spring and dashpot from a platform which is obliged to perform simple harmonic motion of amplitude C and period $2\pi/\omega$ (Fig. 3.8). Let q be the depth of the mass

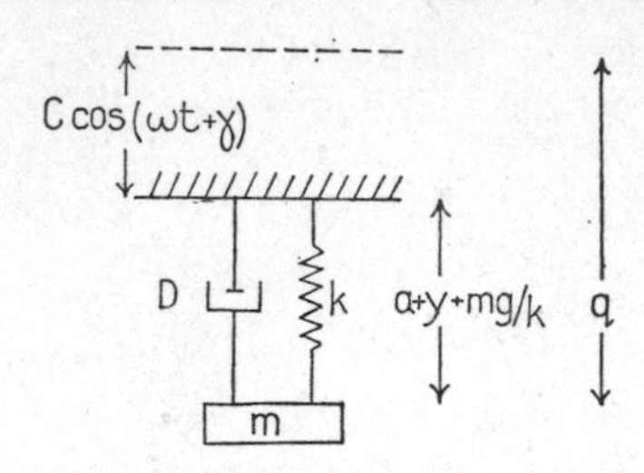

Fig. 3.8
The vibration recorder or seismograph

below the mean position of the platform. The length of the spring is then $q - C\cos(\omega t+\gamma)$. The plunger of the dashpot is moving down with velocity $\frac{d}{dt}C\cos(\omega t+\gamma)$ and the pot with velocity $\dot{q}$. The motion of the pot is therefore opposed by a force $D\{\dot{q}+\omega C\sin(\omega t+\gamma)\}$. Hence $m\ddot{q}=mg-k\{q-a-C\cos(\omega t+\gamma)\}-D\{\dot{q}+\omega C\sin(\omega t+\gamma)\}$.

Now q gives the displacement of the mass as observed in space but we are interested in the displacement as seen on a scale attached to the platform and moving with it. Such a displacement is given by y in the substitution

$$q=C\cos(\omega t+\gamma)+a+mg/k+y.$$

The differential equation for y is

$$m\ddot{y}+D\dot{y}+ky=m\omega^2C\cos(\omega t+\gamma).$$

This is the same as equation (6) with F_0 replaced by $m\omega^2C$ and β by γ. Hence, in the forced oscillation of y,

$$y=Y_0\cos(\omega t+\gamma-\phi)$$

where

$$Y_0=\frac{(\omega^2/\Omega^2)C}{\left\{\left(1-\frac{\omega^2}{\Omega^2}\right)^2+\frac{4b^2\omega^2}{\Omega^2}\right\}^{1/2}}$$

and ϕ is given by (8).

It is easy to check that Y_0/C is the same as X_0k/F with ω/Ω replaced by Ω/ω. Consequently the behaviour of Y_0 can be deduced immediately from Fig. 3.5. Two limiting cases are of interest. As $\omega/\Omega \to 0$, $Y_0/C \to \omega^2/\Omega^2$. In fact, if $b^2 = \frac{1}{2}$, i.e. $b \approx 0{\cdot}7$, $Y_0/C = \omega^2/\Omega^2$ with an error of the sixth order. Here Y_0 provides a measure of the *acceleration* of the platform which is accurate to 3 per cent for $\omega < \frac{1}{2}\Omega$ if $b \approx 0{\cdot}7$.

As $\omega/\Omega \to \infty$, $Y_0 \to C$ so that Y_0 gives a measure of the *amplitude* of oscillation of the platform. From a physical point of view the oscillations are so rapid that the mass stays fixed in space while the platform moves up and down beside it.

Amplitude measurement requires the natural frequency to be less than half the frequency to be recorded whereas acceleration measurement needs the natural frequency to be at least double the frequency being recorded.

When several frequencies are present the waveform traced by the mass may bear little resemblance to the waveshape followed by the platform. For, although the amplitudes of the separate waves are reproduced correctly, the phases do not in general bear the same relationship to one another as in the original waveshape. For example, if the platform follows 2 sine waves, the forced oscillation of the mass will be

$$A_1 \cos(\omega_1 t + \gamma_1 - \phi_1) + A_2 \cos(\omega_2 t + \gamma_2 - \phi_2).$$

On replacing t by $t + \phi_1/\omega_1$ we see that the phase relations are different from the original unless $\phi_2 = \phi_1 \omega_2/\omega_1$. In general ϕ does not depend linearly on ω and this equation cannot be satisfied. However, if $b = 1/\sqrt{2}$, $\phi = \sqrt{2}\omega/\Omega$ correct to 3 per cent for $\omega/\Omega < \frac{1}{2}$. Hence, by choosing $b \approx 0{\cdot}7$ accelerations can be reproduced with little phase distortion if all frequencies present are less than half the natural frequency. We have already seen that there is not much amplitude distortion under the same conditions.

3.8. Operational solution. To complete the investigation we find the general solution for given initial conditions and an arbitrary applied force. The equation of motion can be written

$$\ddot{x}+2b\Omega\dot{x}+\Omega^2x=f(t).$$

Let $x=x_0$, $\dot{x}=x_1$ at $t=0$. Let $F(p)$ be the operational representation of $f(t)$. Then (**RII 2.1** and **2.2**)

$$(p^2+2b\Omega p+\Omega^2)x=F(p)+(p^2+2b\Omega p)x_0+px_1.$$

Now

$$\frac{1}{p^2+2b\Omega p+\Omega^2}=\frac{1}{(p+b\Omega)^2+n^2}$$
$$=\frac{1}{2in}\left\{\frac{1}{p+b\Omega-in}-\frac{1}{p+b\Omega+in}\right\}$$

and $\dfrac{F(p)}{p-\lambda}$ is the representation of $\int_0^t e^{\lambda(t-\tau)}f(\tau)d\tau$. Hence

$$x=e^{-b\Omega t}\left\{x_0\cos nt+\frac{x_1+b\Omega x_0}{n}\sin nt\right\}$$
$$+\frac{1}{n}\int_0^t f(\tau)\,e^{-b\Omega(t-\tau)}\sin n(t-\tau)d\tau. \tag{12}$$

This is the most convenient form when n is real. If n is purely imaginary it is better to convert the sine of imaginary argument into the sinh of real argument by means of $\sin iz=i\sinh z$.

When $f(t)=(F_0/m)\cos(\omega t+\beta)$, (12) reduces to

$$x=e^{-b\Omega t}\left\{\xi_0\cos nt+\frac{\xi_1+b\Omega\,\xi_0}{n}\sin nt\right\}+X_0\cos(\omega t+\beta-\phi)$$

where

$$\xi_0=x_0-X_0\cos\phi,$$
$$\xi_1=x_1-\omega X_0\sin\phi.$$

One can deduce at once that, if x_0 and x_1/ω are of the same order as X_0, the transient can be neglected with less than 10 per cent error if $e^{-b\Omega t}<\frac{1}{10}$, i.e. if $t>2{\cdot}3/b\Omega$. Thus the transient can be ignored after one natural period if $b>0{\cdot}4$. Large damping reduces the transient response but also makes resonance less pronounced.

EXERCISES ON CHAPTER III

1. In the system of Fig. 3.2 the mass is 10 lb. and the extension of the spring under the weight of the particle $\frac{1}{2}''$. In free oscillations the amplitude is observed to decay from $0{\cdot}2''$ to $0{\cdot}1''$ in 35 cycles. Find the frequency of the damped oscillations. Calculate the coefficient of damping; what is its ratio to critical damping?

2. A light rod is smoothly pivoted at one end and carries a particle of mass m at a distance l from the pivot. A vertical spring of stiffness k is attached to a point distant $3l$ from the pivot. The rod is in equilibrium in the horizontal position. A vertical dashpot, with coefficient of damping D, is connected at a point $2l$ from the pivot. Find the frequency of damped oscillations.

3. A solid steel shaft, of torsional stiffness 1,000 ft.-poundals and clamped at one end, carries a disc of moment of inertia 1,000 ft.-poundals. sec.^{2}. The damping torque is $D\dot{\phi}$ where ϕ is the angle of twist and D is 6,000 ft.-poundals. sec. The disc is twisted through an angle of $0{\cdot}31$ radians and then released from rest. How long is it before the angle of twist is $0{\cdot}2$ radians?

4. A microphone consists of a movable vertical plate, of mass m, separated by an air gap from a fixed vertical plate. The mounting of the movable plate resists motion by a force $2mc \times$ (velocity) and restores the plate to its equilibrium position by a force $5mc^2 \times$(deflection), both forces being horizontal and applied to the centre of mass. The output of the microphone is undistorted provided that the separation between the plates is not less than $\alpha d (0 < \alpha < 1)$ where d is the equilibrium separation. Sound waves exert a horizontal force $F_0 \cos \omega t$ on the centre of mass of the movable plate. Show that the output due to the forced oscillation is undistorted for all ω if $2c(1 - \alpha)^{1/2} \geqslant (F_0/md)^{1/2}$.

5. In a standard system of mass supported by a spring and dashpot from a fixed platform it is found that the forced oscillation has a maximum amplitude of X when the frequency of the applied force is $\omega_0/2\pi$. At forcing frequencies of $\omega_0(1 \pm \epsilon)/2\pi$ where $\epsilon << 1$ the amplitudes X_1, X_2 of the forced oscillation satisfy $X_1 = X_2 << X$. Find the ratio of the coefficient of damping to critical damping.

6. A simple pendulum of mass m is such that, when its point of suspension O is fixed and air resistance neglected, its period of small oscillations is $2\pi/\Omega$. O is now attached to one end of a spring of stiffness $3m\Omega^2$ which can slide in a smooth horizontal groove, the other end of the spring being fixed. Air resistance exerts a horizontal force $m\Omega \times$(velocity) on the bob. If a small horizontal force $F_0 \cos \omega t$ is applied to the bob, show that the amplitude of the forced oscillations of the bob is a maximum when $\omega = \frac{1}{2}\Omega$.

7. As a crude model of bicycling over a rough road consider a light circular wheel rolling with constant horizontal velocity v along the sinusoidal road $y = a \cos (x/d)$, the x- and y-axes being horizontal and vertically upwards respectively. The action of the tyres is simulated by a vertical spring of stiffness k and dashpot with damping coefficient D both joined to the lowest point of the wheel and carrying the mass m above them. Show that the amplitude of the forced oscillations relative to the wheel has a maximum of $2kma/D(4km - D^2)^{1/2}$.

Find the amplitude of the force transmitted to the road and show that it pays to have soft springing. What is the advantage of the damping? Show that as the speed increases the force on the road increases as v^2 for a rigid wheel ($k = \infty$) but as v for a tyred wheel.

8. A particle of mass m is suspended from a platform by a spring of stiffness $m\Omega^2$; also connected between the platform and particle is a dashpot with coefficient of damping $2mb\Omega$. The platform is made to oscillate vertically with period $2\pi/\omega$ and amplitude a. Find the amplitude A of the forced oscillations of the particle *in space*. For what value of ω is this amplitude a maximum? If b is small show that this maximum is approximately $a/2b$.

Prove also that A is independent of the damping if $\omega = \sqrt{2}\Omega$ and that an increase of damping increases A if $\omega > \sqrt{2}\Omega$.

9. The mounting of an electric motor is equivalent to a spring of stiffness 230,400 poundals/ft. and a dashpot with coefficient of damping 20 per cent of critical. The motor is subject to an alternating force with the same frequency as that of rotation. If the mass of the motor is 1,000 lb. find the range of rotational speeds for which the force transmitted to the foundation is (*a*) greater than the exciting force, and (*b*) less than 20 per cent of the exciting force.

CHAPTER FOUR

Electrical Circuits

4.1. The fundamental circuit elements. It is well known that when a steady current I flows in a wire there is a relation between the potential difference V across the wire and the current, namely

$$V = RI. \tag{1}$$

Here R is a constant independent of I and known as the *resistance* of the wire. Equation (1) is usually known as *Ohm's law*. Any device, whether it be a wire or not, which satisfies (1) is known as a *resistor* and represented diagrammatically as in Fig. 4.1.

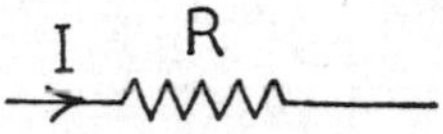

Fig. 4.1
The resistor

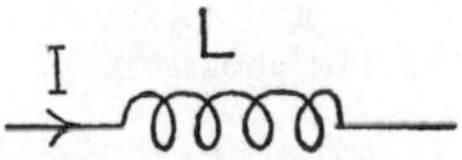

Fig. 4.2
The inductor

If the current is changing in time (1) is still valid with the instantaneous values of V and I substituted but there may be a relation between V and I even if $R = 0$. For example, it is found that if the wire is coiled there is the relation

$$V = L\frac{dI}{dt} \tag{2}$$

when $R=0$. L is a constant known as the *inductance* of the circuit. (The potential difference is induced by the changing magnetic field of the current.) Any device which satisfies an equation of the type (2) is known as an *inductor* and represented diagrammatically as in Fig. 4.2.

The final circuit element to be considered is the ***capacitor***. In simple form a capacitor consists of two parallel plates close together; therefore the pictorial representation of Fig. 4.3 is employed. If Q is the charge on the capacitor the potential difference is given by

$$V=Q/C \tag{3}$$

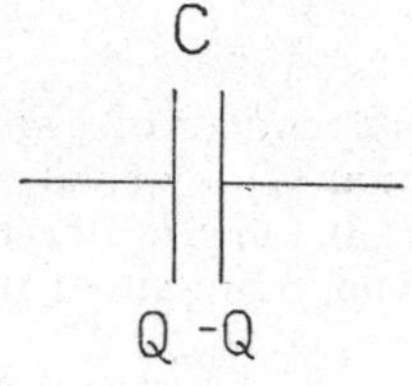

FIG. 4.3
The capacitor

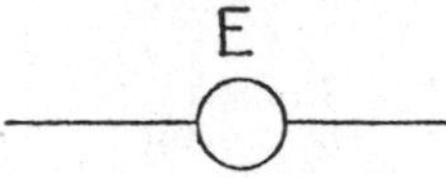

FIG. 4.4
The generator

where the constant C is called the *capacitance*. If the charge is varying in time there is the further relation

$$I=dQ/dt. \tag{4}$$

An electrical source or *generator* will be shown diagrammatically as in Fig. 4.4. This indicates that the generator produces an *electromotive force* (E.M.F., for short) of E, i.e. the potential difference across the generator is obliged to be E.

The only electrical systems to be considered in this chapter are those which can be represented in terms of inductors, resistors and capacitors. All non-linear devices such as electric arcs, valves (unless operating on a linear part of the characteristic) or coils with saturated iron cores are excluded. Furthermore, it will be assumed that the current is the same at all points of the circuit, i.e. *the current is function of time but not of position.* This restriction avoids the partial differential equations necessary for propagation along a coaxial cable. The theory is often called *lumped circuit* theory.

There is a number of different systems of units in use. We shall describe only the practical (or M.K.S.) system of units. The unit of current is the *ampere*, which deposits silver at a certain rate when passed through an aqueous solution of silver nitrate. The unit of charge is the *coulomb* which equals 1 amp. sec. Resistance is measured in *ohms*, one ohm being the resistance of a certain column of mercury. Potential difference is measured in *volts*: 1 volt = 1 amp.-ohm. The unit of capacitance is the *farad* = 1 coulomb/volt and of inductance is the *henry* or 1 volt-sec./amp. The power developed by sending a current I amps. across a potential difference V volts is VI watts.

4.2. Kirchhoff's laws. The rules which enable one to calculate the current flowing in an electrical circuit are:

(i) Current cannot accumulate at a point, i.e. the current flowing into a junction is equal to the current flowing out, and

(ii) The potential difference round a closed circuit is zero.

These rules are usually known as *Kirchhoff's laws.* For the moment we shall be concerned only with the second of them.

Consider, for example, the circuit shown in Fig. 4.5.

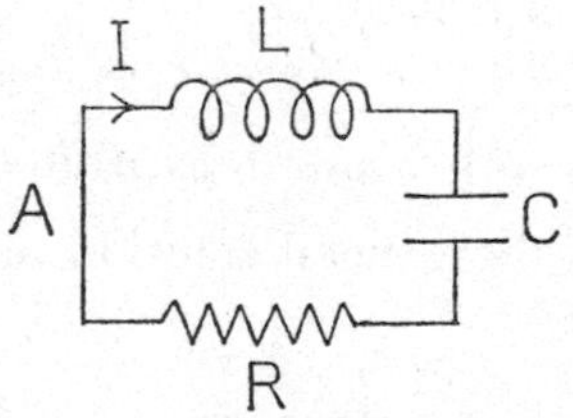

I L
A C
R

FIG. 4.5
The series circuit

The elements are said to be arranged in *series*. Starting from point A and going round the circuit until we return to A we calculate the total potential difference by adding together the separate potential differences as given by (1), (2) and (3). By (ii) the total must be zero. Hence

$$L\frac{dI}{dt}+\frac{Q}{C}+RI=0. \tag{5}$$

Substitute for I from (4); then

$$L\frac{d^2Q}{dt^2}+R\frac{dQ}{dt}+\frac{Q}{C}=0. \tag{6}$$

Alternatively we can take a derivative of (5) and then substitute for dQ/dt from (4), obtaining thereby

$$L\frac{d^2I}{dt^2}+R\frac{dI}{dt}+\frac{I}{C}=0. \tag{7}$$

Both (6) and (7) are of the same type as (III.2) and can be made exactly the same by the identification

$$\Omega^2=\frac{1}{LC},\ 2b\Omega=\frac{R}{L}.$$

All the theory developed in section 3.2 may now be carried over. For example, if $R=0$ both the current and charge vary simple harmonically with frequency $1/2\pi\sqrt{LC}$. There are no oscillations if $b^2\geqslant 1$, i.e. $R^2\geqslant 4L/C$. In particular, if $L=0$,

$$Q=Q_0\,e^{-t/RC}$$

which gives the time constant $1/RC$ for the discharge of a capacitor through a resistor.

If $R^2<4L/C$ there is a damped oscillation of frequency $n/2\pi=\frac{1}{4\pi L}\left(\frac{4L}{C}-R^2\right)^{1/2}$ and logarithmic decrement $\pi R/nL$. On using the general solution of section 3.8 with $X_0=0$ we find

$$Q=Q_0e^{-\frac{1}{2}\frac{Rt}{L}}\left\{\cos nt+\frac{R}{2Ln}\sin nt\right\}$$

when $Q=Q_0$, $\dot{Q}=0$ at $t=0$. The consequent current

$$I=dQ/dt=-\frac{Q_0}{LCn}e^{-\frac{1}{2}\frac{Rt}{L}}\sin nt.$$

When a generator is added to the series circuit, as in Fig. 4.6, the potential difference across the generator must be taken into account in our calculations and

$$L\frac{dI}{dt}+\frac{Q}{C}+RI=E.$$

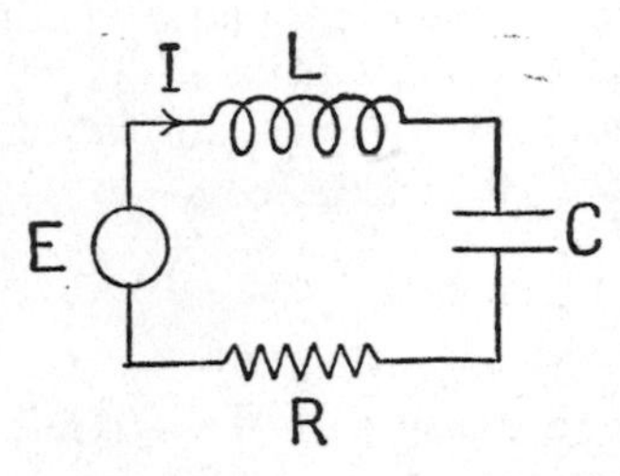

FIG. 4.6
The driven series circuit

If $E=E_0 \cos(\omega t+\beta)$ we derive as before

$$L\frac{d^2Q}{dt^2}+R\frac{dQ}{dt}+\frac{Q}{C}=E_0 \cos(\omega t+\beta),$$

$$L\frac{d^2I}{dt^2}+R\frac{dI}{dt}+\frac{I}{C}=-\omega E_0 \sin(\omega t+\beta).$$

Both of these equations can be discussed as in Section 3.3. In the forced oscillation

$$Q=\frac{E_0 \cos(\omega t+\beta-\phi)}{\left\{\left(\frac{1}{C}-L\omega^2\right)^2+R^2\omega^2\right\}^{1/2}},$$

$$I=\frac{-\omega E_0 \sin(\omega t+\beta-\phi)}{\left\{\left(\frac{1}{C}-L\omega^2\right)+R^2\omega^2\right\}^{1/2}},$$

where

$$\cos\phi = \frac{1 - LC\omega^2}{\{(1 - LC\omega^2)^2 + R^2C^2\omega^2\}^{1/2}},$$

$$\sin\phi = \frac{CR\omega}{\{(1 - LC\omega^2)^2 + R^2C^2\omega^2\}^{1/2}},$$

The amplitude of the charge oscillation is a maximum when $LC\omega^2 = 1 - \frac{1}{2}R^2C/L$ but not the amplitude of the current oscillation. In fact, division of the numerator and denominator by ω reveals that the current is a maximum when $\omega^2 = 1/LC$. There are thus two resonant frequencies of a series circuit depending upon whether one chooses the charge or current. It is usual to place more emphasis on the current and so the resonant frequency of a series circuit is $1/2\pi\sqrt{LC}$ the same as the *natural frequency*. At resonance $\phi = \frac{1}{2}\pi$ and $I = \frac{E_0}{R}\cos(\omega t + \beta)$. At high and low frequencies the current is small.

The current is in phase with the applied E.M.F. at resonance but, in general, there is a phase lag of $\phi - \frac{1}{2}\pi$. The quantity $\sin\phi$ is often known as the *power factor*.

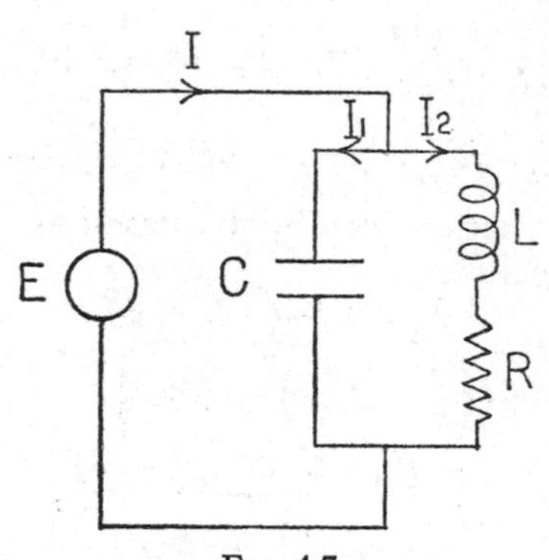

FIG. 4.7
The parallel circuit

The average rate of production of heat is

$$\frac{\frac{1}{2}RC^2E_0^2\omega^2}{(1-LC\omega^2)^2+\omega^2R^2C^2} \text{ watts}$$

which has a maximum of $\frac{1}{2}E_0^2/R$ at resonance.

As an illustration of the use of both of Kirchhoff's laws consider the *parallel circuit* of Fig. 4.7. By law (i)

$$I=I_1+I_2 \tag{8}$$

and by law (ii)

$$\frac{Q}{C}=E, \tag{9}$$

$$L\frac{dI_1}{dt}+RI_1=E \tag{10}$$

where

$$\frac{dQ}{dt}=I_2. \tag{11}$$

From (9) and (11), $I_2/C=dE/dt$
and therefore $I_1=I-CdE/dt$
from (8). Substitution in (10) gives

$$L\frac{dI}{dt}+RI=E+RC\frac{dE}{dt}+LC\frac{d^2E}{dt^2}.$$

If $E=E_0 \cos \omega t$ we derive in the standard way the forced oscillation

$$I=\mathscr{R}\frac{1+i\omega RC-\omega^2LC}{R+i\omega L}E_0\, e^{i\omega t}. \tag{12}$$

In this case the amplitude of the forced oscillation is

$$\left\{\frac{(1-\omega^2LC)^2+\omega^2R^2C^2}{R^2+\omega^2L^2}\right\}^{1/2}.$$

The behaviour of this amplitude is quite different from those considered earlier. The amplitude possesses a ***minimum***, at

$$\omega^2LC=\left(1+\frac{2CR^2}{L}\right)^{1/2}-\frac{CR^2}{L}$$

provided that $1+2CR^2/L>C^2R^4/L$, and no maximum. For $CR^2 << L$, the minimum occurs at $\omega^2LC=1$.

4.3. The electro-mechanical analogy. We have already had one example of an electrical circuit which is governed by exactly the same equation as a certain mechanical system. In general, because of (1) to (4), we can always pass from an electrical circuit to a mechanical system or vice versa if the identification below is made:

Electrical	*Mechanical*
Inductance L	Mass m
Resistance R	Coefficient of damping D
Reciprocal of capacitance $1/C$	Stiffness k
Applied E.M.F. E	Applied force F
Capacitor charge Q	Displacement x
Current I $(=\dot{Q})$	Velocity v $(=\dot{x})$

Statements such as 'the force on the spring is kx' or 'the force on the dashpot is Dv' become 'the potential difference across the capacitor is Q/C' or 'the potential difference across the resistor is RI'. For example, Fig. 4.8 shows the electrical circuits equivalent to the mechanical systems of Ex. 2.2.

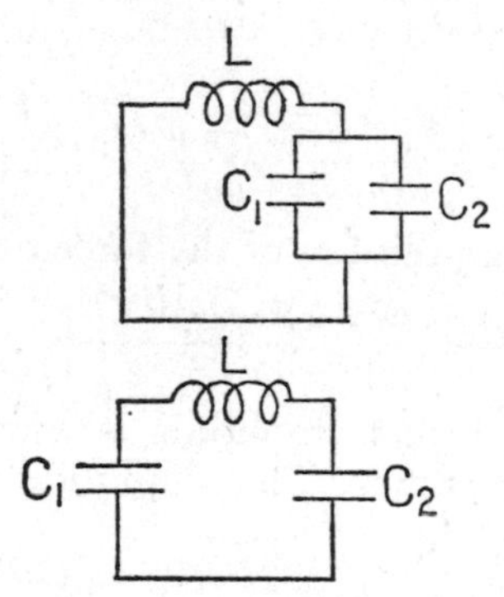

FIG. 4.8
Circuits equivalent to Figs. 2.2 and 2.3

There may be a difference of emphasis in considering the behaviour of two equivalent systems. For instance, most interest usually falls on the displacement in mechanical systems whereas the current is generally of more importance in electrical problems.

The equivalence between electrical and mechanical problems forms the basis of *analogue computers* in which the behaviour of complicated mechanical systems is investigated by examining the conduct of the equivalent and (usually) more easily constructed electrical circuit.

4.4. Complex impedances. In many instances only the forced oscillation is of interest and the transients are of no consequence. As we have seen the forced oscillation can be calculated in a comparatively simple way once the differential equation is known. We now study this point more closely with a view to dispensing with the differential equation.

The equation for the current in the circuit of Fig. 4.6 is

$$L\ddot{I}+R\dot{I}+I/C=-\omega E_0 \sin(\omega t+\beta).$$

The forced oscillation is found by observing that $-\sin(\omega t+\beta)=\mathscr{R}ie^{i(\omega t+\beta)}$ so that a particular solution is $\mathscr{R}\, A\, e^{i(\omega t+\beta)}$. Let $\mathbf{I}=A\, e^{i(\omega t+\beta)}$, $\mathbf{E}=E_0\, e^{i(\omega t+\beta)}$. Then the equation to determine A can be written

$$(-\omega^2 L+i\omega R+1/C)\mathbf{I}=i\omega\mathbf{E}$$

or

$$\mathbf{E}=Z(i\omega)\mathbf{I} \tag{13}$$

where

$$Z(i\omega)=R+i\omega L+1/i\omega C.$$

The current in the forced oscillation is now calculated from $I=\mathscr{R}\mathbf{I}$.

Since $\mathscr{R}\mathbf{E}$ is the applied E.M.F. (13) expresses a generalized form of Ohm's law for simple harmonic currents. $Z(i\omega)$ is called the *complex impedance* and $Y(i\omega)=1/Z(i\omega)$ is the *complex admittance*. The factor of i in the impedance, i.e. $\omega L-1/\omega C$ is known as the *reactance*.

If we work with the complex potential difference $\mathbf{E}$, complex current **I** and complex impedances we can still use Kirchhoff's laws because, if w, w_1 and w_2 are complex numbers, $w=w_1+w_2$ implies that $\mathcal{R}w=\mathcal{R}w_1+\mathcal{R}w_2$.

Now turn to the case in which two series circuits are connected in series (Fig. 4.9). Instead of showing the resistor, inductor and capacitor in each series circuit we have

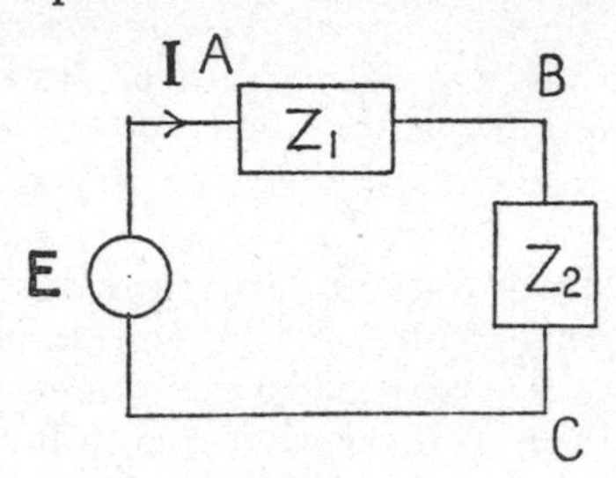

FIG. 4.9
Impedances in series

represented them by the appropriate complex impedances, e.g.

$$Z_1(i\omega)=R_1+i\omega L_1+1/i\omega C_1.$$

Let $I=\mathcal{R}\mathbf{I}$ be the current flowing. Then, by (13) the complex potential difference from A to B is $Z_1\mathbf{I}$ and from B to C is $Z_2\mathbf{I}$. But the complex potential difference from A to C is $\mathbf{E}$. Hence

$$\mathbf{E}=Z_1\mathbf{I}+Z_2\mathbf{I}=Z\mathbf{I}$$

where

$$Z=Z_1+Z_2. \tag{14}$$

The equation connecting **E** and **I** is of the same form as (13). Hence, *two complex impedances in series are equivalent to a complex impedance which is equal to their sum.* In particular, two resistors in series have an equivalent resistance of R_1+R_2, two inductors an equivalent reactance of $\omega(L_1+L_2)$ and two capacitors an equivalent reactance of $1/\omega C_1+1/\omega C_2$. Obviously the rule can be extended to any number of impedances in series.

Consider now two series circuits connected in parallel

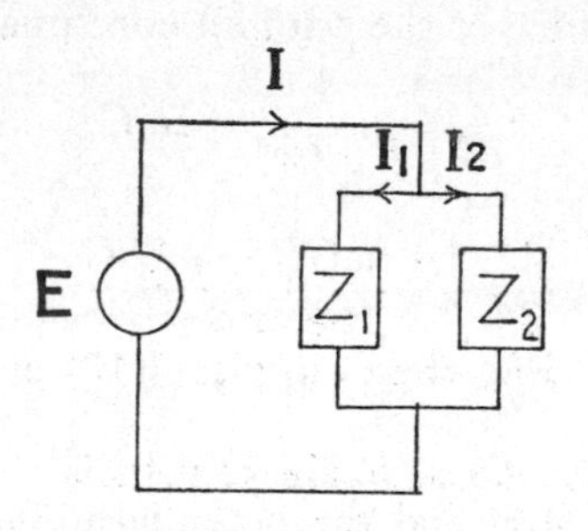

FIG. 4.10
Impedances in parallel

(Fig. 4.10). The complex potential difference across Z_1 is $Z_1\mathbf{I}_1$ and across Z_2 is $Z_2\mathbf{I}_2$. Hence, by Kirchhoff's laws,

$$\mathbf{E}=Z_1\mathbf{I}_1,$$
$$\mathbf{E}=Z_2\mathbf{I}_2,$$
$$\mathbf{I}=\mathbf{I}_1+\mathbf{I}_2.$$

Therefore

$$\mathbf{I}_1=Z_2\mathbf{I}_2/Z_1=Z_2\mathbf{I}/(Z_1+Z_2)$$

and

$$\mathbf{E}=\frac{Z_1\,Z_2}{Z_1+Z_2}\mathbf{I}$$
$$=Z\,\mathbf{I}$$

where

$$\frac{1}{Z}=\frac{1}{Z_1}+\frac{1}{Z_2}. \tag{15}$$

Thus, *two complex impedances in parallel are equivalent to a complex impedance whose reciprocal is the sum of their reciprocals.*

The two rules (14) and (15) permit the calculation of current flow through complicated electrical circuits in a comparatively straightforward manner. For example, in the circuit of Fig. 4.7 the complex impedance of the arm containing the inductor and resistor is $R+i\omega L$ and the complex impedance of the other arm is $1/i\omega C$. Hence the complex impedance of the parallel combination is given by

$$\frac{1}{Z}=\frac{1}{R+i\omega L}+i\omega C$$

or

$$\frac{1}{Z}=\frac{1+i\omega CR-\omega^2 LC}{R+i\omega L}.$$

The formula $\mathbf{I}=\mathbf{E}/Z$ then supplies (12) at once.

Example 4.1. The E.M.F. $E_0 \sin 5t$ volts is connected between A and C of Fig. 4.11 and the potential difference between B and C is required.

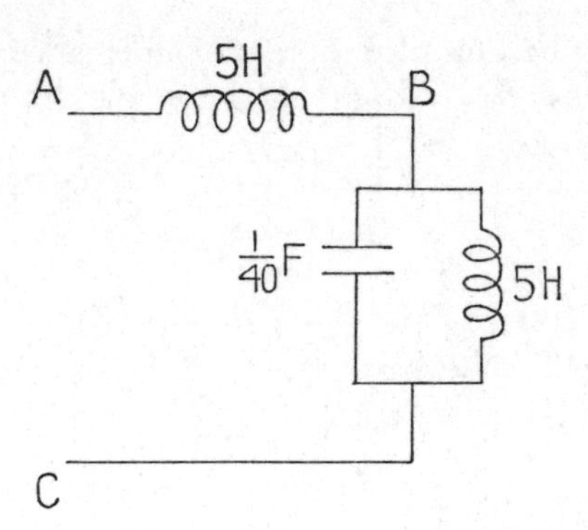

FIG. 4.11

The complex impedance between B and C is, by the computation just made with $R=0$,

$$\frac{i\omega L}{1-\omega^2 LC}=-\frac{200i}{17}$$

on putting $\omega=5$, $L=5$, $C=\frac{1}{40}$. Hence the impedance between A and C is $225i/17$. Since $\mathbf{E}=-i\,E_0\,e^{5it}$,

$$\mathbf{I}=-\frac{17}{225}E_0\,e^{5it}.$$

The complex potential difference between B and C is $-\frac{200}{17}i\,\mathbf{I}$.

Hence the required potential difference is

$$\mathscr{R}\frac{200}{225}iE_0e^{5it}=-\frac{8}{9}E_0\sin 5t.$$

Note: The equivalent mechanical circuit consists of two discs connected by a shaft performing torsional oscillations.

4.5. Operational impedances. The analysis of the preceding section is not of much help if the transients are needed. There is, however, a somewhat similar process which can be adopted to give the transients under suitable conditions.

We start by considering the equations for the series circuit of Fig. 4.6, namely

$$L\frac{dI}{dt}+\frac{Q}{C}+RI=E,$$

$$\frac{dQ}{dt}=I.$$

These equations can be solved by means of p-operators (**RII 2.5**). We have

$$(pL+R)\text{I}+\text{Q}/C=\text{E}+pLI_0,$$

$$p\text{Q}=\text{I}+pQ_0$$

where I, Q, E are the operational representations and $Q=Q_0$, $I=I_0$ at $t=0$. Eliminating Q from these equations we obtain

$$(pL+R+1/pC)\text{I}=\text{E}+pLI_0-Q_0/C \qquad (16)$$

which determines the operational representation of the current.

It may happen that $Q_0=0$ and $I_0=0$; then

$$\text{E}=Z(p)\text{I}$$

where

$$Z(p)=R+pL+1/pC.$$

By analogy with the foregoing $Z(p)$ is called the *operational impedance* of the circuit. The analysis concerning the combination of impedances proceeds in the same fashion as that following (13). A summary of the results is:

(a) Kirchhoff's laws can be applied as if the operational representations of potential difference and current and the operational impedance were ordinary potential difference, current and circuit elements; the operational representation of the potential difference being the product of the operational impedance and operational representation of the current.

(b) Operational impedances in series are equivalent to an operational impedance which is equal to their sum.

(c) Operational impedances in parallel are equivalent to an operational impedance whose reciprocal is the sum of their reciprocals.

(d) The complex impedance is obtained from the operational impedance by putting $p=i\omega$.

Example 4.2. At time $t=0$ the E.M.F. $E_0 \sin 5t$ is applied between A and C (Fig. 4.11). Find the subsequent potential difference between B and C, assuming that the charges and currents are initially zero.

The operational impedance of the capacitor is $40/p$ and of the parallel inductor is $5p$. Hence the operational impedance between B and C is the reciprocal of $\dfrac{1}{5p}+\dfrac{p}{40}=\dfrac{p^2+8}{40p}$, i.e. it is $\dfrac{40p}{p^2+8}$.

Hence the operational impedance between A and C is

$$5p+\frac{40p}{p^2+8}=\frac{5p(p^2+16)}{p^2+8}.$$

The operational representation of the applied E.M.F. is $5E_0p/(p^2+25)$. Hence the equation for the current is

$$\frac{5p(p^2+16)\mathrm{I}}{p^2+8}=\frac{5E_0p}{p^2+25}$$

or

$$\mathrm{I}=\frac{(p^2+8)E_0}{(p^2+16)(p^2+25)}.$$

It is a straightforward matter to interpret this formula but, since we are asked for the potential difference between B and C, we need

$$\frac{40p\,\text{I}}{p^2+8}=\frac{40p\,E_0}{(p^2+16)(p^2+25)}$$
$$=\frac{40}{9}p\,E_0\left\{\frac{1}{p^2+16}-\frac{1}{p^2+25}\right\}$$

whence the required potential difference is

$$\frac{1}{9}E_0\,(10\sin 4t\;-8\sin 5t).$$

The reader should compare the answers of Examples 4.1 and 4.2. They differ in the term involving sin $4t$. This is because the method of complex impedances gives only that part of the answer which has the same sinusoidal variation as the applied E.M.F. The answer of Ex. 4.1 is substantially in error because there is no damping to remove the sin $4t$. In any practical circuit, of course, there would always be some damping present.

Broadly speaking, we can say that if the damping is sufficient to remove all terms except those with the same time variation as the applied E.M.F. the method of complex impedances is adequate. The initial conditions are of no importance. If the transients cannot be neglected use operational impedances provided that the initial charges and currents are zero. For cases which do not fall into either of these categories the full equations involving initial conditions, e.g. (16) must be solved.

EXERCISES ON CHAPTER IV

1. In a series electrical circuit the inductance is 0·25 henries, the resistance 200 ohms and the capacitance 10^{-6} farads. Initially, the potential difference across the capacitor is 100 volts and there is no current. Find the subsequent charge on the capacitor.

2. An electromagnet of inductance 1 henry and negligible resistance carries a steady current of 10 amps. A capacitor is connected across the magnet. What must be its capacitance in order that the maximum potential difference across the magnet is 10,000 volts when the current supply is broken? If this capacitance be used and the electromagnet has a resistance of 1 ohm show that the maximum potential difference is 9,992 volts approximately.

3. A parallel circuit of inductance L, capacitance C and negligible resistance is connected in series with a resistor of resistance R and a battery of constant E.M.F. E. When a steady current has been established in the circuit the E.M.F. is suddenly reduced to zero. Show that the subsequent charge on the capacity is $(E/Rn)e^{-\frac{1}{2}t/RC}\sin nt$ where $n^2 = \dfrac{1}{LC} - \dfrac{1}{4R^2C^2}$.

4. Show that there is no reactance in the circuit of Fig. 4.7 if $\omega = \omega_0$ where $\omega_0^2 = \dfrac{1}{LC} - \dfrac{R^2}{L^2}$. Prove also that the ratio of the amplitude of the current when ω is not near ω_0 to the amplitude when $\omega = \omega_0$ is $\dfrac{\omega}{\omega_0}\left(1 - \dfrac{\omega_0^2}{\omega^2}\right)\sqrt{\dfrac{L}{CR^2}}$ if $CR^2 << L$.

5. A resistor of resistance R is inserted in series with the capacitor in the circuit of Fig. 4.7. Prove that the amplitude of the current is the same at all frequencies if $CR^2 = L$.

6. Show that, in the circuit of Fig. 4.12, the magnitude of the impedance is independent of C and R if $2\omega^2 = 1/LC$. Find the phase at this frequency.

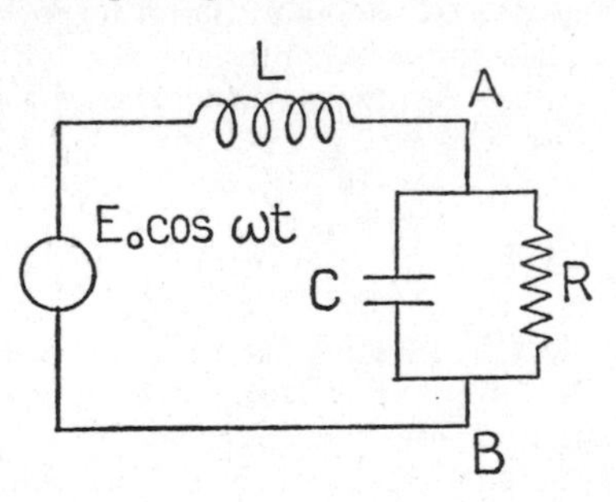

FIG. 4.12

If $\omega^2 = 1/LC$ prove that the complex potential difference between A and B is $-iR\,E_0\,e^{i\omega t}\sqrt{C/L}$ and deduce that the current through the resistor is independent of its resistance.

7. In the circuit of Fig. 4.13 the charges and currents are zero until, at time $t = 0$, an E.M.F. $E_0 \sin \omega t$ is applied between A and B. Find the subsequent potential difference across C_2 if $\omega^2 L(C_1 + C_2) \neq 1$.

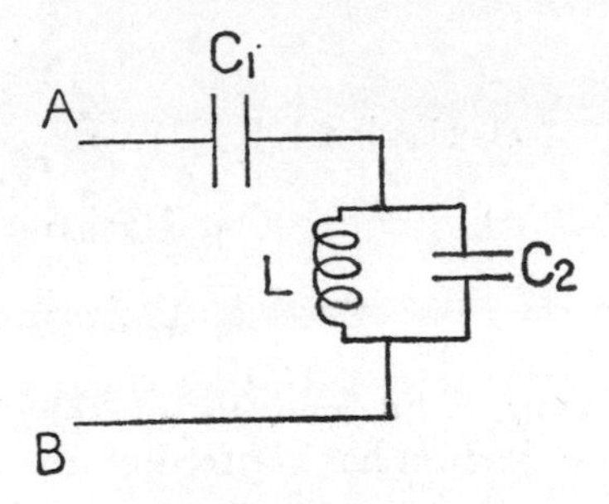

FIG. 4.13

8. The capacitor C_1 in Fig. 4.13 is replaced by a resistor of resistance 4 ohms and $L=5H$, $C_2=\frac{1}{20}F$. The charges and currents are zero until an E.M.F. $E_0 \cos 4t$ is applied between A and B. Calculate the subsequent potential difference across the resistor.

9. In the circuit of Fig. 4.13, $L=1H$, $C_1=\dfrac{1}{50}F$ and C_2 is replaced by a resistor of resistance 5 ohms. The charges and currents are zero until, at $t=0$, an E.M.F. $E(t)$ is applied between A and B. Find the subsequent potential difference across C_1.

Obtain the potential difference in the cases (i) $E(t)=t$, (ii) $E(t)=E_0 \sin 10t$.

CHAPTER FIVE

Non-Linear Oscillations

5.1. Introduction. The analysis of the behaviour of a system near equilibrium has centred on the equation

$$m\ddot{x}+D\dot{x}+kx=0. \tag{1}$$

If $D=0$ there is simple harmonic motion of which the amplitude is determined by the initial conditions. If $D>0$ the only oscillatory motion is damped and eventually dies out. Now there are many important vibrating physical systems which do not possess these features. For example, the amplitude of oscillation of a watch mechanism is the same no matter how it is started and the oscillation persists despite friction at the bearings. An electrical example is a valve oscillator which produces the same oscillation irrespective of the excitation. It may be objected that in both these examples sources are present, namely the mainspring of a watch and the high tension supply of a valve oscillator but this does not resolve the difficulty. Admittedly the addition of a forcing term on the right-hand side of (1) produces a forced oscillation whose amplitude is independent of the initial conditions but the frequency is equal to that of the applied force whereas in both our examples oscillatory motion is generated by constant sources. Therefore our explanation must lie elsewhere.

Let us first of all see what happens to the energy in (1). Let $\mathscr{E}$ be the total energy so that $\mathscr{E}=\frac{1}{2}m\dot{x}^2+\frac{1}{2}kx^2$. Then

$$\dot{\mathscr{E}}=m\dot{x}\ddot{x}+kx\dot{x}=-D\dot{x}^2.$$

Therefore, if $\mathscr{E}=\mathscr{E}_0$ at $t=0$,

$$\mathscr{E}-\mathscr{E}_0=-\int_0^t D\dot{x}^2\,dt. \tag{2}$$

Now, if $D>0$, the integral is positive and increasing. Hence $\mathscr{E}$ is always less than $\mathscr{E}_0$ and decreases steadily until it becomes zero. The energy is being dissipated in the form of heat by the friction.

Suppose now that $D<0$; then $\mathscr{E}$ steadily increases. The possibility now exists of the build-up of oscillations in the absence of a forcing term. Of course, the extra energy must be supplied by the device which is responsible for the *negative damping*. In practice, the infinite energy predicted by (2) as $t \to \infty$ does not occur so that negative damping alone is not sufficient to explain our phenomena. Nevertheless, it suggests a possible mechanism in which the damping is negative to start with, diminishing in magnitude and becoming zero as the system attains its final state, any greater motion being prevented by the damping going positive. Such a mechanism can be produced by (1) if D is a function of x or $\dot{x}$ or both; (1) is then *non-linear*.

Before describing the essential differences between linear and non-linear differential equations we remark that non-linearities can arise through m and k as well as through D. For example, in a rotating Catherine-wheel the mass is continually changing through the discharge of hot gas. If extra springs are brought into action at certain displacement, as is done in some kinds of railway buffers, k depends on x. In the electrical circuit the inductance L will be a function of the current I if the inductor contains an iron core which can be magnetically saturated. The potential difference across a resistor of thyrite is not proportional to I but to $I^{0 \cdot 28}$ and a capacitor containing Rochelle salt has a capacitance which varies with the charge, roughly as $(1+aQ+bQ^2)^{-1}$.

It should be pointed out that m, D and k being variable does not necessarily make (1) non-linear. The equation is still linear if they are all functions of the variable t alone. One or more of m, D and k must be a function of x (and/or its time derivatives) for the equation to be non-linear.

The effect of non-linearity can be conveniently displayed by considering the particular equation

$$\ddot{x}+x^2=0. \tag{3}$$

In it the stiffness, being a quadratic function of x, is non-linear. One characteristic property of a linear equation is that, if $x=X_1(t)$ and $x=X_2(t)$ are solutions, so is X_1+X_2. Non-linear equations lack this property. For the addition of

$$\ddot{X}_1+X_1^2=0,$$
$$\ddot{X}_2+X_2^2=0$$

gives

$$\ddot{X}_1+\ddot{X}_2+(X_1+X_2)^2-2X_1X_2=0.$$

The presence of the term $-2X_1X_2$ shows that X_1+X_2 does not satisfy the differential equation.

Another characteristic of a linear equation is that if $x=X_1$ is a solution so is $x=CX_1$ where C is an arbitrary constant. But substitution of $x=CX_1$ in (3) reveals that it is a solution only in the trivial cases $C=0$ and $C=1$.

Consequently the general method of solving second order linear equations by adding together arbitrary multiples of two independent solutions is not available for non-linear equations. This makes the exact analysis of non-linear equations extremely difficult and, except in rare cases, recourse to approximate methods or numerical integration is usually necessary.

5.2. The phase plane. When employing numerical methods, with or without the assistance of a digital computer, it is usually much simpler to deal with first order differential equations rather than second order. For this reason, we introduce the variable y defined by

$$\dot{x}=y. \tag{4}$$

An equation such as

$$\ddot{x}+x^2\dot{x}+x^3=0 \tag{5}$$

can then be written as

$$\dot{y} = -x^2y - x^3. \tag{6}$$

In other words the single second order equation (5) is replaced by the two first order equations (4) and (6). More generally, we can visualize a more complicated substitution than (4) e.g.

$$\dot{x} = P(x, y) \tag{7}$$

leading to an equation of the form

$$\dot{y} = Q(x, y). \tag{8}$$

The numerical problem is then one of integrating (7) and (8), a straightforward matter normally.

As a matter of fact the conversion to (7) and (8) is often of considerable help analytically. Instead of the general system (7) and (8) we consider only (4) and (8), i.e. we suppose our equations are

$$\dot{x} = y, \quad \dot{y} = Q(x, y).$$

The plane in which x and y are Cartesian co-ordinates is called the *phase plane*. It displays the variation of the velocity y with the displacement x. For a given solution (x, y) describes a curve in the phase plane as the time varies. This curve is known as a *trajectory*. The differential equation of the trajectory can be deduced easily; for

$$\frac{dy}{dx} = \frac{\dot{y}}{\dot{x}} = \frac{Q(x, y)}{y}.$$

It is not always easy to integrate.

Having found a trajectory one wishes to know the direction which (x, y) travels along it. Now, if $y > 0$, $\dot{x} > 0$ and x must be increasing as t increases. Hence, in $y > 0$, (x, y) *travels along the trajectory in the direction which makes x increase*. Similarly, in $y < 0$, the direction is that which makes x decrease.

It can be shown that, under suitable conditions, there is only one trajectory through a given point of the phase plane, i.e. two trajectories do not intersect in general.

5.3. Equilibrium. An equilibrium position has been defined earlier as a configuration at which vanishing velocity implies zero acceleration. In our present notation the velocity is y and the acceleration $\dot{y}$. Hence at equilibrium $y=0$ and x satisfies $Q(x, 0)=0$.

The time required to cover a portion of a trajectory is $\int dt=\int dx/\dot{x}=\int dx/y$. Since y vanishes at equilibrium *any approach to equilibrium by the system occupies an infinite time.*

To illustrate the use of the phase plane in discussing motion about equilibrium let us consider the equation of simple harmonic motion

$$\ddot{x}+\Omega^2 x=0$$

which is equivalent to the system

$$\dot{x}=y,\ \dot{y}=-\Omega^2 x.$$

The trajectories satisfy $\dfrac{dy}{dx}=-\Omega^2 x/y$ and are

$$\Omega^2 x^2+y^2=C^2$$

where C is an arbitrary constant. Plotted in the phase plane the trajectories form a series of ellipses with the origin as common centre. (Fig. 5.1) Each of the ellipses corresponds to a possible motion which can be initiated with suitable

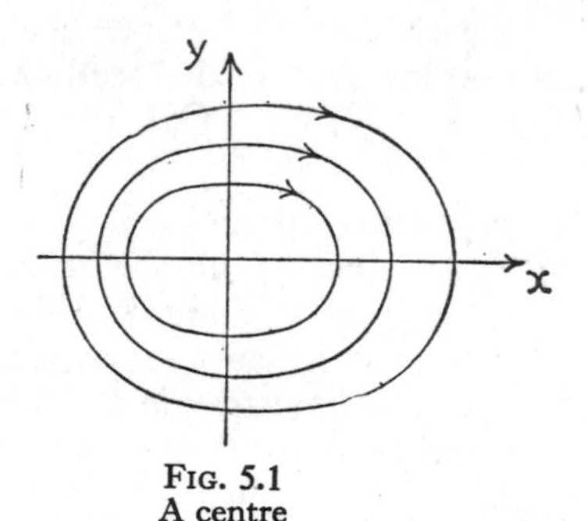

FIG. 5.1
A centre

initial conditions. The arrows show the direction of traversing a trajectory according to the rule at the end of the preceding section.

Although the phase plane does not give the precise variation of x with t a considerable amount of information can be deduced from it. For example, having gone round an ellipse once we return to the original values of x and y. Since this occurs continually the motion is periodic. The maximum displacement from the equilibrium position ($x=0, y=0$) is C/Ω. The period of the motion is $\int dx/y$ taken round the ellipse, i.e.

$$4\int_0^{C/\Omega} \frac{dx}{(C^2-\Omega^2x^2)^{1/2}}=2\pi/\Omega$$

in agreement with earlier results.

Quite generally, we can say that a motion is *periodic* if $x(t+T)=x(t)$ for all t. The period is T if T is the smallest possible number for which this is true. A time derivative gives $y(t+T)=y(t)$. Thus the trajectory of a periodic motion always returns to its starting point. Alternatively, we can state *when a trajectory is a closed curve the corresponding motion is periodic*. It should be emphasized that a periodic motion need not be simple harmonic. An equilibrium position which is surrounded by closed trajectories as in Fig. 5.1 (but not necessarily ellipses) is frequently called a *centre*.

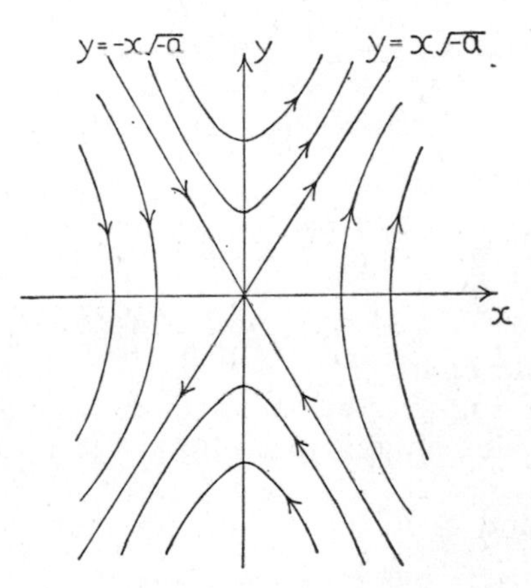

FIG. 5.2
The saddle-point

It is of interest to examine the behaviour of (1) in the phase plane. Write the equation in the form

$$\ddot{x}+2b\dot{x}+ax=0$$

so that $\dot{x}=y$, $\dot{y}=-ax-2by$. The case $b=0$, $a>0$ has already been discussed. Now suppose $b=0$, $a<0$; the trajectories are the hyperbolae $y^2+ax^2=C$. (Fig. 5.2) Whatever the initial conditions (unless they represent points on the line $y=-x\sqrt{-a}$) the system eventually moves away from the equilibrium position which is therefore unstable. If the initial state is such that $y=-x\sqrt{-a}$ the system will move towards equilibrium but only reach there after an infinite time. An equilibrium position as depicted in Fig. 5.2 is known as a *saddle-point*.

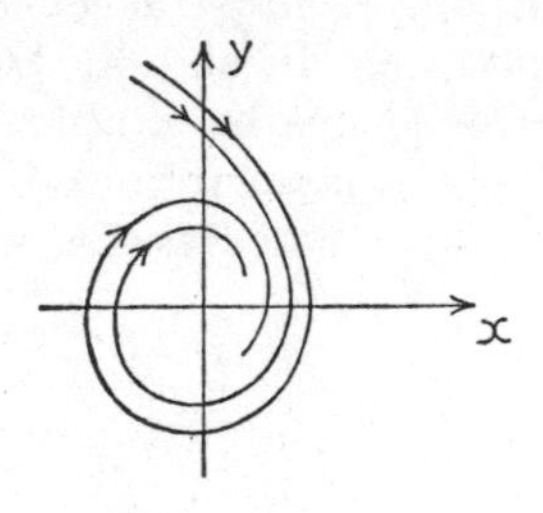

FIG. 5.3
The focus

If $b\neq 0$ and $a<0$ it can be verified that the trajectories are

$$(y+c_1x)^{c_1}(y+c_2x)^{c_2}=C$$

where c_1 and c_2 are $|b\pm\sqrt{b^2-a}|$. The curves are similar to those of Fig. 5.2 but rotated through an angle about the origin. The equilibrium position is a saddle-point and unstable.

If $a>0$ and $a-b^2=n^2>0$ the equation of the trajectories is

$$y^2+2bxy+ax^2=C\,e^{\frac{2b}{n}\tan^{-1}\frac{y+bx}{nx}}\ .$$

The curves are spirals which spiral into the origin if $b>0$ (Fig. 5.3).
The motion is a damped oscillation and the equilibrium position is stable. It is known as a *focus*. If $b<0$ the curves spiral out of the origin (the term focus is still employed) and the system moves away from equilibrium with oscillations of ever increasing amplitude. The position is unstable.

If $a>0$ and $b^2 - a \geqslant 0$ the trajectories are

$$(y+\lambda_1 x)^{\lambda_1}=C(y+\lambda_2 x)^{\lambda_2}$$

where $\lambda_1=b-\sqrt{b^2-a}$, $\lambda_2=b+\sqrt{b^2-a}$. If $b>0$ the disposition is shown in Fig. 5.4. Any motion tends to

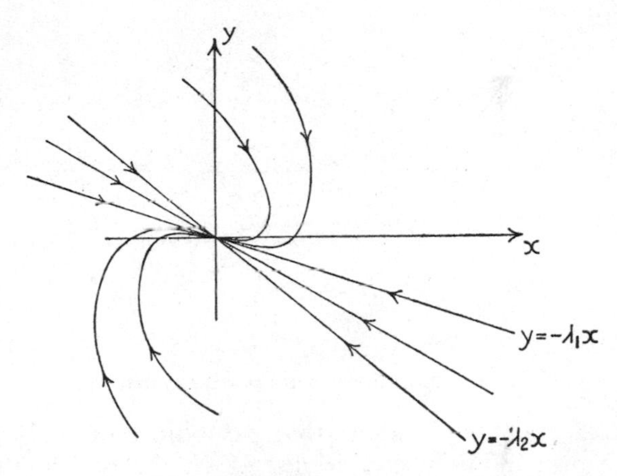

FIG. 5.4
The node

equilibrium and the position is stable. It is known as a *node*. If $b<0$ the structure is similar but the motion is always away from equilibrium so that the position is unstable.

As another example of the application of the phase plane consider the equation

$$\ddot{x}+\tfrac{1}{2}(x^2-1)=0$$

which is a non-linear conservative system. The equations for x and y are

$$\dot{x}=y,\ \dot{y}=\tfrac{1}{2}(1-x^2)$$

whence

$$y\frac{dy}{dx}=\tfrac{1}{2}(1-x^2).$$

Integration gives the trajectories

$$y^2=x-\tfrac{1}{3}x^3+C.$$

These are shown graphically in Fig. 5.5. The point $x=1$ is a centre and $x=-1$ a saddle-point. If $C<\frac{2}{3}$ the closed

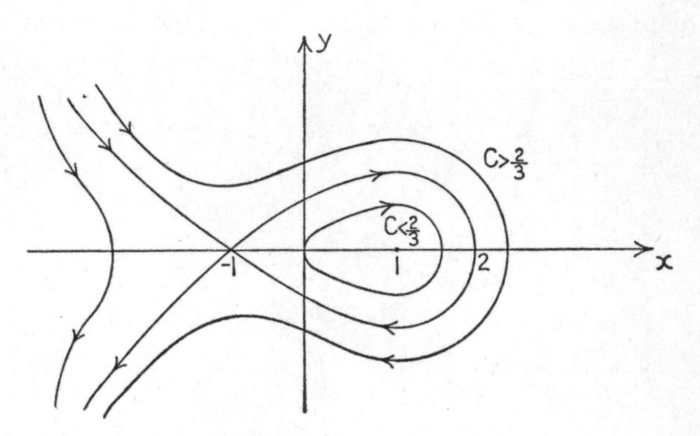

Fig. 5.5
Non-linear conservative system

curves about $x=1$ show that periodic motion is possible with proper initial conditions. If $C=\frac{2}{3}$ no oscillations are permitted although the system can approach $x=-1$ steadily if the initial conditions are suitable. If $C>\frac{2}{3}$ the system always ends up by going to infinity in the direction of negative x.

Non-linear *conservative* systems can always be dealt with in this way. The equation of the trajectories represents the conservation of energy and is

$$\tfrac{1}{2}y^2+V(x)=C$$

where V is the potential energy. Some readers will find it helpful to think of the trajectories as the projections on the phase plane of the contours of the surface $z=\frac{1}{2}y^2+V(x)$, the z-axis be perpendicular to the (x, y)-plane.

5.4. Slightly non-linear damping. We now turn our attention to non-conservative systems in which a small amount of damping is present. Such a system might be governed by an equation such as

$$\ddot{x}+\epsilon f(x, \dot{x})+x=0$$

where $\epsilon<<1$. To fix ideas, we consider the specific example of a simple valve oscillator. The reader who is unfamiliar with the properties of valves is advised to jump to equation (9) and accept that this derivation is valid in appropriate circumstances.

In Fig. 5.6 it is assumed that the inductor in the anode circuit is coupled to the inductor L in such a way as to produce an E.M.F. MdI_a/dt in the grid circuit. Here I_a is

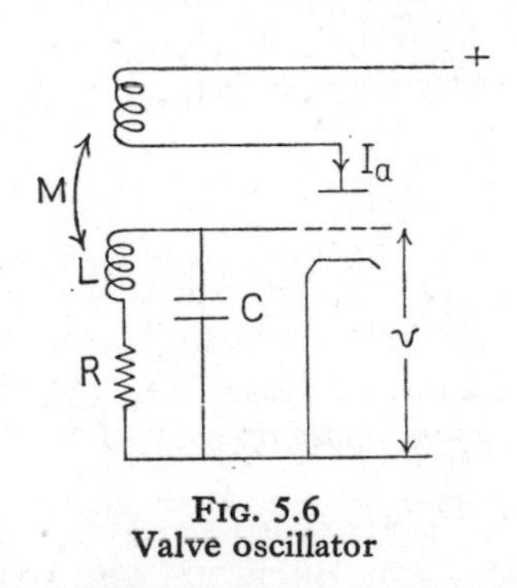

FIG. 5.6
Valve oscillator

the anode current. Further it is assumed that there is no current flow through the grid. The current through L is then the same as that through C. If Q is the charge on the capacitor

$$\dot{Q}=I,$$
$$L\dot{I}+RI+Q/C=M\dot{I}_a.$$

Eliminating I we obtain

$$L\ddot{Q}+R\dot{Q}+Q/C=M\dot{I}_a.$$

The grid-cathode potential difference v is given by $v=Q/C$. There is a relation between v and I_a which is expressed graphically in Fig. 5.7.

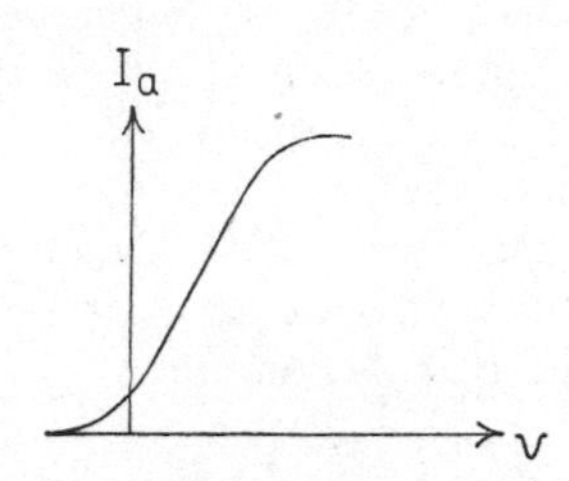

Fig. 5.7
Relation between anode current and grid voltage

Write $I_a=g(v)$. Then the equation for Q is

$$L\ddot{Q}+\left\{R-\frac{M}{C}g'\left(\frac{Q}{C}\right)\right\}\dot{Q}+Q/C=0 \qquad (9)$$

where $g'(v)=dg(v)/dv$.

By our general criterion of section 5.1 oscillations can be expected to build up only if there is negative damping for small values of Q. This requires $CR<Mg'(0)$, a condition we shall assume to be satisfied from now on.

It is obvious that few results can be anticipated for general values of g so that we shall now introduce the approximation

$$g(v)=\alpha+\beta v-\gamma v^3$$

where $\beta>0$ and $\gamma>0$. Such an approximation may be expected to be adequate for moderate values of v but will be badly in error for large $|v|$. Nevertheless it should reproduce the main features of the effect of the bending of the anode-grid characteristic in Fig. 5.7. Equation (9) becomes

$$L\ddot{Q}+\{R-M\beta/C+3M\gamma Q^2/C^3\}\dot{Q}+Q/C=0$$

and the necessary condition for oscillations is $R<M\beta/C$. Make the charge of variables

$$t=\tau/\Omega,\ \Omega^2=1/LC,\ Q=x\{C^2(M\beta-CR)/3M\gamma\}^{1/2}$$

and the equation reduces to

$$x''-\epsilon(1-x^2)x'+x=0 \tag{10}$$

where $x'=dx/d\tau$ and $\epsilon=(M\beta-CR)\Omega$.

Equation (10) is known as *van der Pol's* equation.

In (10) ϵ is a non-negative constant which can be made to take any value by adjusting the circuit constants. At the moment, however, we are concerned only with small values of ϵ. One's first idea is to ignore the term involving ϵ—the equation then predicts simple harmonic motion. This suggests that we try $x=A\cos\tau/T$ where T is nearly equal to 1. Substitution in (10) gives

$$A(1-1/T^2)\cos\tau/T=-\epsilon(1-A^2\cos^2\tau/T)(A/T)\sin\tau/T$$
$$=(\epsilon A/T)\{\sin 3\tau/T-(1-\tfrac{1}{4}A^2)\sin\tau/T\}.$$

This equation will be satisfied for all τ only if the coefficients of the different sinusoidal terms vanish. The best we can do is to choose

$$T=1,\ A=2.$$

The term $\sin 3\tau/T$ remains unaccounted for. To remove that we must try $x=A\cos\tau/T+B_1\cos 2\tau/T+C_1\sin 2\tau/T$ where T is nearly 1, A is nearly 2, B_1 and C_1 are small. Clearly, however we choose T, A, B_1 and C_1 there will be additional terms in the equation which can only be removed by adding terms involving $3\tau/T$, $4\tau/T$, . . . to the expression for x. The process can be terminated as soon as the quantities are known to a sufficiently high power of ϵ. For most purposes the information supplied by our simple substitution is enough. It tells us that there is a periodic motion

$$x=2\cos\tau$$

with an error of order ϵ in amplitude and an error of order ϵ^2 in period.

The introduction of even a small amount of non-linear damping modifies radically the motion. Instead of periodic motion with any amplitude being possible, as in the simple harmonic case, only one amplitude is permitted.

It has not been proved that this periodic state can ever be attained. In fact, this state is reached whatever the initial conditions as will now be demonstrated. Multiply (10) by x' and integrate with respect to τ from τ_1 to τ_2; then

$$\tfrac{1}{2}[x'^2+x^2]_2-\tfrac{1}{2}[x'^2+x^2]_1=\int_{\tau_1}^{\tau_2}\epsilon(1-x^2)x'^2d\tau.$$

The equation states that the change in energy of the system from τ_1 to τ_2 is given by the integral on the right. Now, providing that the interval $\tau_2-\tau_1$ is not too long, we can expect x to be substantially of the form $B\cos\tau$. If we use this formula in the integral the resulting energy change will be in error by a term of order ϵ^2. Let $\tau_2=\tau_1+2\pi$. We have

$$\int_{\tau_1}^{\tau_2}\epsilon(1-x^2)x'^2d\tau=\epsilon\int_{\tau_1}^{\tau_1+2\pi}(1-B^2\cos^2\tau)B^2\sin^2\tau d\tau$$
$$=\tfrac{1}{4}\pi\epsilon B^2(4-B^2).$$

Thus if $B>2$ the energy is reduced after one period. Since the argument can be repeated for each consecutive period we see that if the motion starts with $B>2$ it must end up with $B=2$. Similarly if $B<2$ the energy is increased after each period and continues to do so until $B=2$. Hence the system ends up in periodic motion however it be started.

Curves showing the phase plane and the variation of x with τ are shown in Figs. 5.8 and 5.9 for a certain starting point. A closed trajectory which is eventually reached by a system is called a *limit cycle*, i.e. $x^2+y^2=4$ is a limit cycle in Fig. 5.8.

We shall now give a brief discussion of how the above technique is applied to the general equation

$$x''+x=-\epsilon f(x,x'). \qquad (11)$$

Put $x=A\cos\tau/T$ and expand the right hand side in terms

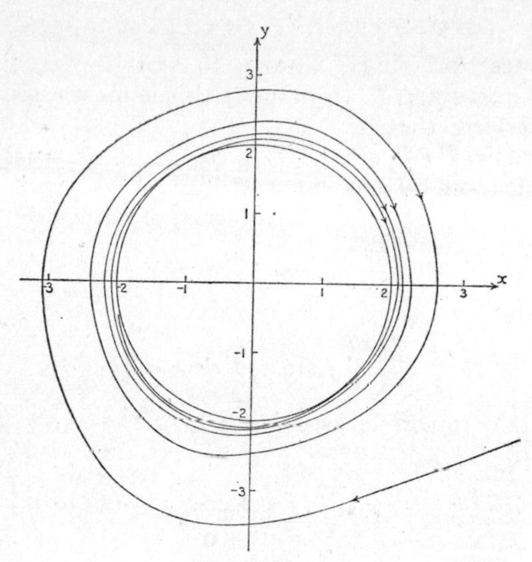

Fig. 5.8
The limit cycle

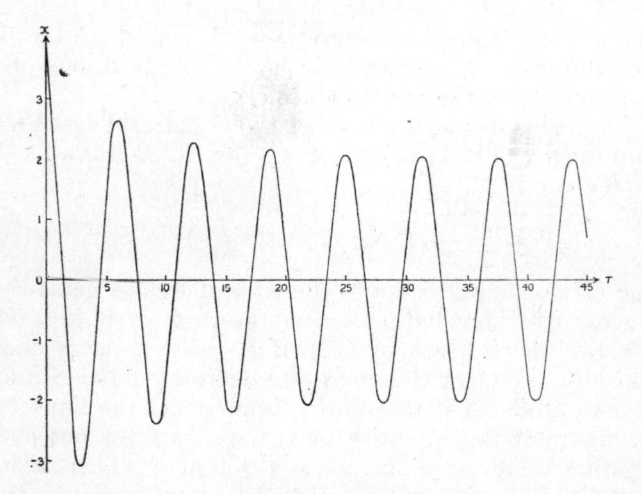

Fig. 5.9
Variation of x with τ

of cosines and sines. Equate to zero the coefficients of $\cos \tau/T$ and $\sin \tau/T$. In practice this is most easily achieved by observing that, if

$f=a_1 \cos \tau/T+b_1 \sin \tau/T+a_2 \cos 2\tau/T+b_2 \sin 2\tau/T+\ldots$, multiplication by $\cos \tau/T$ and integration from 0 to $2\pi T$ gives

$$\int_0^{2\pi T} f \cos \tau/T \, d\tau=\pi T a_1.$$

Similarly

$$\int_0^{2\pi T} f \sin \tau/T \, d\tau=\pi T b_1.$$

Hence the coefficients of $\cos \tau/T$ and $\sin \tau/T$ in (11) vanish if

$$1-1/T^2=-\frac{\epsilon}{A}\int_0^{2\pi T} f\left(A \cos\frac{\tau}{T}, -\frac{A}{T}\sin\frac{\tau}{T}\right)\cos\frac{\tau}{T}\, d\tau$$

and

$$0=\int_0^{2\pi T} f\left(A \cos\frac{\tau}{T}, -\frac{A}{T}\sin\frac{\tau}{T}\right)\sin\frac{\tau}{T}\, d\tau. \qquad (12)$$

The second equation determines A and then the first equation gives T correct to order ϵ. The positions of the limit cycles are thereby established.

It remains to decide whether the system approaches the limit cycle. The change of energy in a period when $x=B \cos \tau$ is

$$+\epsilon B\int_0^{2\pi} f(B \cos \tau, -B \sin \tau)\sin \tau \, d\tau. \qquad (13)$$

One cannot hope to assess the sign of this in general but one can ask what happens near the limit cycle, i.e. when $B \approx A$, A being given by (12). If $B=A$ the energy change vanishes. We want the energy to decrease if $B>A$ and to increase if $B<A$ if the motion approaches the limit cycle i.e. we want the quantity in (13) to go from positive to negative values as B increases through A. This requires that the derivative of (13) shall be negative when $B=A$, i.e.

$$\epsilon\int_0^{2\pi}(f+Af_x\cos\tau-Af_y\sin\tau)\sin\tau\,d\tau<0$$

where $f_x=\frac{\partial}{\partial x}f(x, y), f_y=\frac{\partial}{\partial y}f(x, y)$ and x and y are put equal to $A\cos\tau$ and $-A\sin\tau$ respectively after the derivatives have been performed. Now, by integration by parts,

$$\int_0^{2\pi} f\sin\tau\,d\tau=\int_0^{2\pi}\cos\tau\frac{\partial f}{\partial\tau}d\tau$$
$$=-\int_0^{2\pi}A\cos\tau\{f_x\sin\tau+f_y\cos\tau\}d\tau.$$

Hence the inequality becomes

$$\int_0^{2\pi} f_y\,d\tau>0 \tag{14}$$

since $\epsilon>0$ and $A>0$. This is the condition under which the system tends to go into the periodic motion of the limit cycle, i.e. the limit cycle is *stable*. If the sign $>$ were replaced by $<$ the system would tend to depart from the limit cycle.

The number of possible limit cycles is determined by the number of distinct values of A which satisfy (12). Their stability can be examined by means of (14). With this information it is possible to decide which limit cycle will be attained with given initial conditions.

There is an important theorem available if $f(x, x')=x'g(x)$. Let $g(x)$ be an *even* function of x such that $g(x)<0$ for $|x|<1$, $g(x)>0$ for $|x|>1$. Further, suppose there is an $l>0$ such that $g(x)\geqslant l$ for $|x|\geqslant x_0$. Then the system has just one limit cycle and this limit cycle is stable. This theorem is valid even if ϵ is not small.

We remark that

$$\ddot{z}+F(\dot{z})+z=0$$

can be thrown into this form. For, take a time derivative and put $\dot{z}=x$; then

$$\ddot{x}+F'(x)\dot{x}+x=0.$$

5.5. Relaxation oscillations. The behaviour of systems in which non-linear effects predominate is much more complicated. They are usually characterized by periodic motion in which intervals where slow changes occur alternate with intervals of very rapid variation. The term *relaxation oscillation* is used to describe such periodic motion.

Although it would be more appropriate, from the point of view of practical application, to discuss the multivibrator or the transitron, for simplicity we shall consider van der Pol's equation (10) when ϵ is large. The equation for the trajectories of

$$\ddot{x} - \epsilon(1 - x^2)\dot{x} + x = 0$$

is

$$\frac{dy}{dx} = \epsilon(1 - x^2) - \frac{x}{y}. \qquad (15)$$

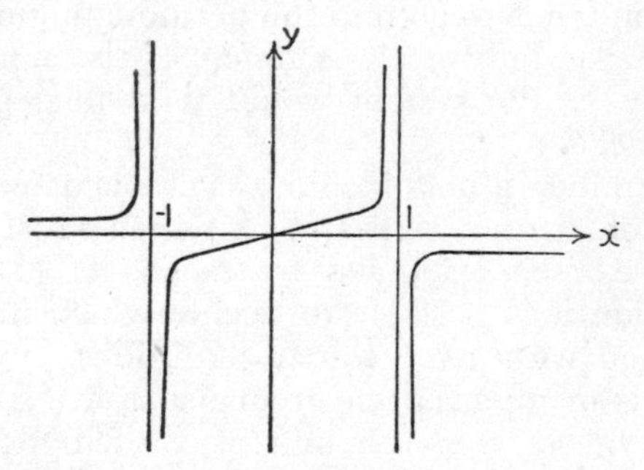

FIG. 5.10
Curves where the trajectory slope is zero

Let the initial conditions be $x = x_0(>1)$, $y = 0$. For very small y (15) is effectively $dy/dx = -x/y$ whence $x^2 + y^2 =$ constant. Therefore the trajectory starts off as the arc of a circle heading in the direction of negative y with decreasing x. As soon as $y<0$ both terms of (15) are large and of the opposite sign. Fig. 5.10 shows the curve $y = x/\epsilon\,(1 - x^2)$ where the slope of the trajectory is zero. Thus almost immediately $\dfrac{dy}{dx}$ goes from a large positive value to a small positive value and the trajectory is clamped between the x-axis and the curve of zero slope. Consequently y diminishes very slightly while x decreases from x_0 to 1. By the time $x=1+O(1/\epsilon)$ the first term in (15) is negligible because $y=O(1/\epsilon)$. The magnitude of y remains small as x passes through 1 in spite of the slope being comparatively large because the interval involved is small. Thereafter the first term dominates the scene.

If the second term in (15) is neglected

$$y=\epsilon(x-\tfrac{1}{3}x^3)+\text{constant}.$$

In view of the foregoing little error will be committed if this be assumed to be valid up to $x=1$. Therefore, if we assume that the trajectory has been traced from $(x_0, 0)$ to $(1, y_0)$ the trajectory from $(1, y_0)$ is

$$y=\epsilon(x-\tfrac{1}{3}x^3-\tfrac{2}{3})+y_0. \tag{16}$$

This formula will still be valid as we pass through $x=-1$ for, although the first term of (15) vanishes at $x=-1$, the second term is only $-\frac{3}{4\epsilon}$ so that only a small error is produced in taking dy/dx zero there.

According to (16) the point $(-2, y_0)$ is on the trajectory and the slope is large and negative. Since y_0 is small the influence of the second term in (15) must be considered. It is negative so that its effect will be to steepen the slope of the trajectory. Since the trajectory is already virtually vertical the additional steepening will be barely visible and the trajectory will pass through $(-2, 0)$.

Now note that if x and y be changed into $-x$ and $-y$ respectively (15) is unaltered. Therefore, the shape of the trajectory from $(-2, 0)$ can be deduced from the preceding. It will creep along near the x-axis from $x=-2$ to $x=-1$, then sweep up to a maximum at $x=1$ and finally fall to the point $(2, 0)$. Having returned to conditions similar to those we started from, the subsequent trajectory presents no problem.

Once again there is a limit cycle (Fig. 5.11) and periodic motion but the behaviour is far from simple harmonic. This is brought out clearly in Fig. 5.12 where the variation of x with t is displayed. Both of these diagrams were obtained by numerical integration.

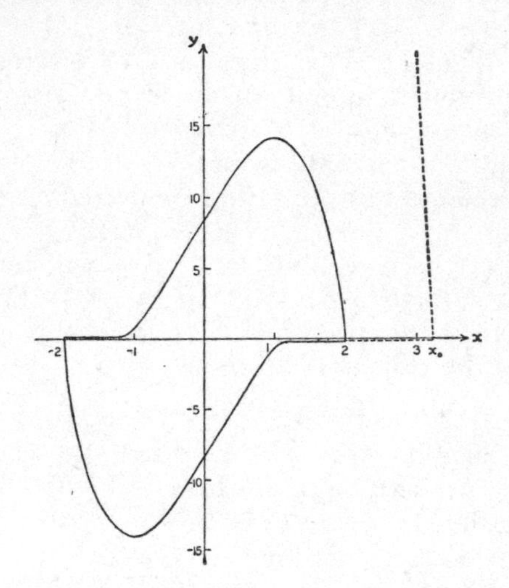

FIG. 5.11
The dotted line shows where the trajectory differs from the limit cycle

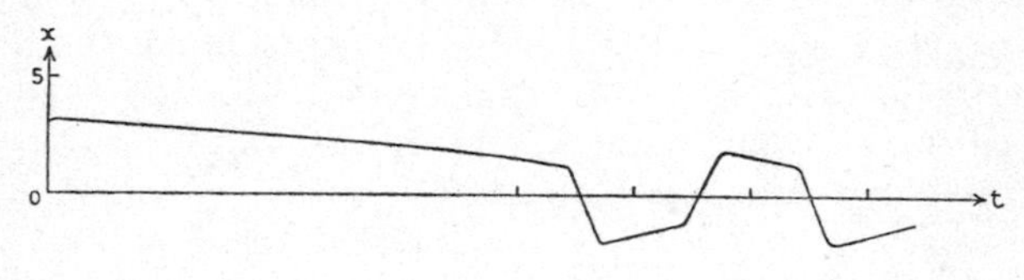

FIG. 5.12
Relaxation oscillation

The general theorem quoted at the end of section 5.4 tells us that there is a limit cycle and that the system arrives there whatever the initial conditions. We have shown this to be true for $x_0 > 1$; we now demonstrate it for other initial conditions. Firstly suppose $0 \leqslant x = x_0 (\leqslant 1)$, $y=0$ initially. Then the trajectory is

$$y = \epsilon(x - x_0 - \tfrac{1}{3}x^3 + \tfrac{1}{3}x_0^3)$$

at the beginning. The trajectory crosses the x-axis again at $2x = -x_0 - \sqrt{3(4 - x_0^2)}$. This point always lies between -2 and $-\sqrt{3}$ so that the subsequent trajectory can be dealt with as before: the limit cycle is joined shortly after crossing the x-axis. For the arbitrary initial conditions $x = x_0$, $y = y_0$ where y_0 is not small the cubic approximation can again be employed. The cubic curve must cross the x-axis at one point at least. When it does so we may proceed as before.

With regard to the period note that in $y>0$ y (or $\dot{x}$) is small from $x = -2$ to $x = -1$ and large for most of the remainder. Hence the system will occupy most of its time from $x = -2$ to $x = -1$ (from $x=2$ to $x=1$ in $y<0$). The precise variation of y in this region is not known but it is certain that y lies below the curve of zero slope $y = x/\epsilon(1 - x^2)$. Hence $\int_{-2}^{-1} dx/y$, the time taken, is greater than $\int_{-2}^{-1} \epsilon(1 - x^2)dx/x$ or $\epsilon\left(\frac{3}{2} - ln2\right)$. Consequently the period is greater than $1{\cdot}6\epsilon$. The contribution of the steeply sloping parts will be negligible by comparison. Thus $1{\cdot}6\epsilon$ can be taken as a crude estimate of the period, the error being due to overestimating y in $-2 \leqslant x \leqslant -1$. In fact, the numerical results show that the estimate is about 25 per cent out, the true period being 2ϵ. It is a surprising but very useful result that integration along the curve of zero slope should provide such a guide to the period.

We conclude this section with a brief discussion of the action of a brake on a moving surface. (The squeaking of chalk on a blackboard, the action of a violin bow and the singing of glass when rubbed by a wet finger all lead to similar mathematics.) As a simple model we consider a mass m, supported by a spring and dashpot, in contact

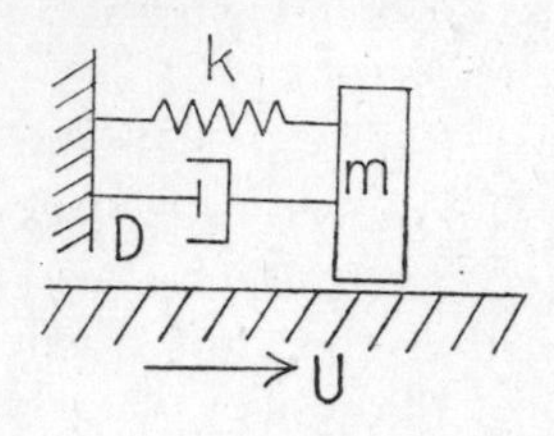

FIG. 5.13

with a surface moving with constant speed U (Fig. 513). Dry friction will act and its variation with relative velocity must be taken into account. The equation of motion of the mass is

$$m\ddot{x}+D\dot{x}+kx=F(U-\dot{x}).$$

The frictional force $F(v)$ decreases as the relative velocity v increases from zero. Also $F(-v)=-F(v)$ because the frictional force changes direction on a reversal of relative velocity. At a first application of the brake $\dot{x}$ will be small and $F(U-\dot{x})\approx F(U)-\dot{x}F'(U)$. The damping will be negative if $D+F'(U)<0$, which may occur for sufficiently low U because F decreases most rapidly near the origin. For large U the damping is usually positive. Thus oscillations are most likely to arise for small U. This explains why chattering or squealing of brakes commonly occurs at low speeds but not at high.

The substitution $t=D\tau/k$ gives

$$k(\mu x''+x'+x)=F(U-kx'/D)$$

where $\mu=mk/D^2$. The equation of the trajectories is

$$\mu y\frac{dy}{dx}=G(y)-x \qquad (17)$$

where $kG(y)=F(U-ky/D)-ky$. A typical form for G is given by the curve $x=G(y)$ in Fig. 5.14.

In many cases the mass is small so that μ is small. Then (17) indicates that y' is large and y changing rapidly unless it is on $x=G(y)$. Hence a trajectory tends to go in a vertical direction until it strikes the curve $x=G(y)$. It then tends to follow this curve until prevented by a bend. For example

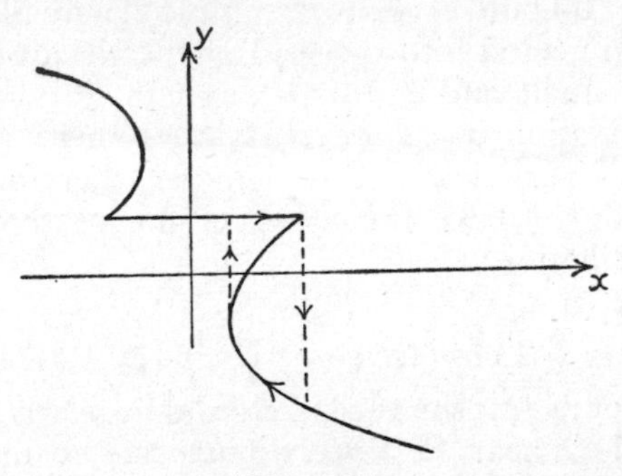

FIG. 5.14
Limit cycle of a brake

suppose a point on the portion parallel to the x-axis is reached. We then move along this line to the right until the corner is attained. Since the trajectory cannot go in the direction of negative x it must move off and therefore into the region of large slope. It drops almost vertically until the curve $x=G(y)$ is again met. This curve is followed until once again the trajectory becomes vertical. Thus we have a limit cycle as shown by the arrows in Fig. 5.14.

The physical explanation given by Rayleigh is that the wheel always moves faster than the brake so that the frictional force is always in the same direction. When the brake moves in the same direction as the wheel the friction is large and assists motion whereas when the brake is receding the friction is small and retards motion. Thus the energy is increased over a cycle and the oscillation can build up.

5.6. Forced oscillations. The analysis of the effect of an applied force on a non-linear system is extremely difficult in general. If the non-linearity is small some insight can be obtained by the method of section 5.4. As an illustration of the technique consider

$$\ddot{x} - \epsilon(1 - x^2)\dot{x} + x = E \sin \omega t. \qquad (18)$$

This is van der Pol's equation with an applied simple harmonic force. It could arise, for example, if a small alternating E.M.F. were injected into the grid of the circuit of Fig. 5.6. When ϵ is small and $E=0$ the system goes into a self-excited oscillation $x=2\cos t$. If $\epsilon=0$ there is a forced oscillation of period $2\pi/\omega$. In general, therefore, when ϵ is small and $E\neq 0$ we must expect the presence of both types of oscillation.

Therefore, try

$$x=A\cos(\Omega t+\gamma)+C\sin(\omega t+\theta) \tag{19}$$

where, except near resonance, C should be nearly $E/(1-\omega^2)$, θ small and Ω near 1. No estimate can be made of the values of A and γ. The phases γ and θ have been introduced to account for the phase difference of the forced oscillation from the applied force and the phase difference between the free and forced oscillations.

The substitution of (19) into (18) gives

$$\begin{aligned}
&A(1-\Omega^2)\cos(\Omega t+\gamma)+C(1-\omega^2)\sin(\omega t+\theta)-E\sin\omega t\\
&\quad=\epsilon[\omega C(1-\tfrac{1}{2}A^2-\tfrac{1}{4}C^2)\cos(\omega t+\theta)+\\
&\qquad(\tfrac{1}{4}A^2+\tfrac{1}{2}C^2-1)\Omega A\sin(\Omega t+\gamma)\\
&-\tfrac{1}{4}A^2C(\omega+2\Omega)\cos\{(\omega+2\Omega)t+2\gamma+\theta\}+\\
&\quad\tfrac{1}{4}A^3\Omega\sin 3(\Omega t+\gamma)\\
&+\tfrac{1}{4}A^2C(2\Omega-\omega)\cos\{(\omega-2\Omega)t-2\gamma+\theta\}+\tfrac{1}{4}\omega C^3\cos 3(\omega t+\theta)\\
&-\tfrac{1}{4}AC^2(\Omega+2\omega)\sin\{(\Omega+2\omega)t+\gamma+2\theta\}\\
&+\tfrac{1}{4}AC^2(2\omega-\Omega)\sin\{(\Omega-2\omega)t+\gamma-2\theta\}].
\end{aligned} \tag{20}$$

Unless $\omega=0$, $\tfrac{1}{3}\Omega$, Ω or 3Ω all the sinusoidal terms on the right-hand side of (20) have different arguments. Let us leave on one side these exceptional cases ($\omega=0$ has already been dealt with). Then, equating the coefficients of $\cos(\Omega t+\gamma)$, $\sin(\Omega t+\gamma)$, $\sin\omega t$ and $\cos\omega t$ to zero, we obtain

$$A(1-\Omega^2)=0, \tag{21}$$

$$(\tfrac{1}{4}A^2+\tfrac{1}{2}C^2-1)\Omega A=0, \tag{22}$$

$$C(1-\omega^2)\cos\theta+\epsilon\omega C(1-\tfrac{1}{2}A^2-\tfrac{1}{4}C^2)\sin\theta=E, \tag{23}$$

$$C(1-\omega^2)\sin\theta-\epsilon\omega C(1-\tfrac{1}{2}A^2-\tfrac{1}{4}C^2)\cos\theta=0. \tag{24}$$

From (23) and (24) it follows that
$C=E/(1-\omega^2)$, $\tan\theta=\epsilon\omega C(1-\frac{1}{2}A^2-\frac{1}{4}C^2)/(1-\omega^2)$.
Thus the non-linearity has very little influence on the forced oscillation. From (21) and (22)

$$\Omega=1,\ A=\sqrt{4-2C^2}.$$

As C increases from 0, A decreases from 2 to 0 and finally becomes imaginary for $C>\sqrt{2}$. Since imaginary values are not permitted $A=0$ is the only possible solution to (21) and (22) when $C>\sqrt{2}$. The effect of the forcing term, therefore, is to reduce the amplitude of the self-excited oscillation if $C<\sqrt{2}$ and *to extinguish it completely* if $C\geqslant\sqrt{2}$, i.e. $E\geqslant\sqrt{2}(1-\omega^2)$.

Turning to the exceptional cases let us first assume that $\omega=3\Omega$. Since Ω is near 1 this can occur only when $\omega=3+\delta$ where $\delta<<1$ so that $\Omega=1+\frac{1}{3}\delta$. The terms involving $\sin 3(\Omega t+\gamma)$ and $\cos\{(\omega-2\Omega)t-2\gamma+\theta\}$ in (20) must now be taken into account. Equating to zero the coefficients of $\cos(\Omega t+\gamma)$, $\sin(\Omega t+\gamma)$, $\sin 3\Omega t$ and $\cos 3\Omega t$ we have

$$A(1-\Omega^2)+\tfrac{1}{4}\epsilon A^2C\Omega\cos(3\gamma-\theta)=0, \tag{25}$$

$$(\tfrac{1}{4}A^2+\tfrac{1}{2}C^2-1)\Omega A-\tfrac{1}{4}\epsilon A^2C\Omega\sin(3\gamma-\theta)=0, \tag{26}$$

$$C(1-9\Omega^2)\cos\theta-E+3\epsilon\Omega C(1-\tfrac{1}{2}A^2-\tfrac{1}{4}C^2)\sin\theta =\tfrac{1}{4}\epsilon A^3\Omega\cos 3\gamma, \tag{27}$$

$$C(1-9\Omega^2)\sin\theta-3\epsilon\Omega C(1-\tfrac{1}{2}A^2-\tfrac{1}{4}C^2)\cos\theta =\tfrac{1}{4}\epsilon A^3\Omega\sin 3\gamma. \tag{28}$$

From (27) and (28) $C=-\frac{1}{8}E$ and $\tan\theta$ is small so that the forced oscillation is virtually the same as in the absence of the non-linearity. In (25) and (26) put $\Omega=1+\frac{1}{3}\delta$, $A\cos(3\gamma-\theta)=\xi$, $A\sin(3\gamma-\theta)=\eta$. Then

$$-\tfrac{2}{3}\delta+\tfrac{1}{4}\epsilon C\xi=0$$

and

$$\xi^2+(\eta-\tfrac{1}{2}C)^2=4-\frac{7}{4}C^2.$$

Thus ξ and η (and thereby A and γ) are determined by the intersection of a straight line and circle. The circle is imaginary if $C^2 > 16/7$ and the line does not intersect the circle if $4 - \frac{7}{4}C^2 < (8\delta/3\epsilon C)^2$. Therefore A is non-zero only if $4 - 7C^2/4 > (8\delta/3\epsilon C)^2$ and then

$$A^2 = 4 - \frac{3}{2}C^2 \pm \left\{4C^2 - \frac{7}{4}C^4 - \left(\frac{8\delta}{3\epsilon}\right)^2\right\}^{1/2}. \qquad (29)$$

The amplitude A having been determined, the quantity γ can be found. However, γ occurs only in the combination 3γ so that if γ_0 is a possible value so are $\gamma_0 + \frac{2}{3}\pi$ and $\gamma_0 + \frac{4}{3}\pi$. In other words for a given amplitude of oscillation of period $2\pi/\Omega$ there correspond three possible distinct phases.

The present investigation does not allow us to say which of the many oscillations will be attained by the system starting from given initial conditions. A more elaborate analysis reveals that the negative square root in (29) corresponds to an unstable state whereas the positive square root corresponds to a stable state. Therefore, under conditions in which A exists, the system tends to adopt the larger value of A; the phase will depend on the initial conditions.

As C^2 increases A will start at some value, increase to a maximum and then decrease to zero. Once again there is extinction of the self-excited oscillation by the forced but it requires a larger value of E than before. However, the increase of A and the possibility of its being greater than 2 is a new feature. For instance, if $\delta = 0$, $A \geqslant 2$ for $0 \leqslant C^2 \leqslant 1$. In general $A \geqslant 2$ if $3\epsilon > 8\delta$ and

$$-\{1 - (8\delta/3\epsilon)^2\}^{1/2} \leqslant 2C^2 - 1 \leqslant \{1 - (8\delta/3\epsilon)^2\}^{1/2}.$$

This magnification of the self-excited oscillation at an integral sub-multiple of the exciting frequency is known as *subharmonic resonance*.

The exceptional case $\omega = \frac{1}{3}\Omega$ may be discussed in a similar manner. In the case $\omega = \Omega$ both terms in (19) are no longer necessary and we may put $C=0$. The phenomenon of resonance still occurs.

The same technique may be employed for equations which have small non-linearities in the spring stiffness instead of in the damping term. No attempt will be made here to formulate a theory for the general equation.

More information about non-linear equations will be found in A. A. ANDRONOW and C. E. CHAIKIN, *Theory of Oscillations*, Princeton (1949); J. J. STOKER, *Non-linear Vibrations, Interscience* (1950).

EXERCISES ON CHAPTER V

1. Draw the trajectories in the phase plane of the differential equations

(i) $\ddot{x}-k\dot{x}=0$,

(ii) $\ddot{x}=8x\dot{x}$,

(iii) $\ddot{x}+3\dot{x}+2x=0$,

(iv) $\ddot{x}-4\dot{x}+40x=0$ by means of the substitution $x=\rho\cos\phi$, $y=\rho\sin\phi$. Discuss the motion of the system.

2. A capacitor and inductor are connected in series. The capacitance is $1/(2+3aQ+4bQ^2)$ when the charge is Q. Examine the possibility of a periodic variation of Q with time.

3. If a particle slides on a parabolic wire rotating with constand angular velocity ω about a vertical axis the distance x of the particle from the axis of rotation satisfies

$$(1+a^2x^2)\ddot{x}+(ag-\omega^2+a^2\dot{x}^2)x=0.$$

Analyse the possible types of motion.

4. The bob of a simple pendulum of length l and mass m is subject to a horizontal force $m\omega^2 l\sin\theta$ tending to increase θ, the inclination to the vertical. Show that the equation of motion is

$$\ddot{\theta}=\omega^2(\cos\theta-\mu)\sin\theta$$

where $\mu=g/\omega^2 l$. Sketch the trajectories for $-\pi\leqslant\theta\leqslant\pi$.

5. The potential difference across a special resistor is RI^2 where R is a positive constant and I the current. The resistor is connected in series with inductance L and capacitance $1/L$. Form the equation for the charge Q on the capacitor and, by

making the substitution $Q^2 = w$, show that periodic variation of the charge is possible.

6. The displacement x of a spring-mounted mass under the action of Coulomb dry friction satisfies

$$m\ddot{x} + kx = \begin{cases} -F \ (\dot{x} > 0) \\ F \ (\dot{x} < 0) \end{cases}$$

where m, k and F are positive constants. If $x = x_0$, $\dot{x} = 0$ at $t = 0$ where $x_0 > 3F/k$ show that after the first cycle the value of x is $x_0 - 4F/k$.

If, when $x = -a$, the kinetic energy is increased by $\mathscr{E}$ show that periodic motion is possible provided that $\mathscr{E} > 8F^2/k$, the largest amplitude occurring being $\frac{F}{k} + \frac{1}{4}\frac{\mathscr{E}}{F}$. (This forms a possible model for a clock mechanism.)

7. In the oscillator of Fig. 5.6 the valve characteristic is given by

$$g(v) = \alpha + \beta v + \delta v^2 - \gamma v^3.$$

Show that, with suitable substitutions, the equation for the charge on the capacitor can be written

$$x'' - \epsilon(1 + \eta x - x^2)x' + x = 0$$

where $\eta = 2M\delta/\sqrt{3\gamma M(M\beta - RC)}$. If both ϵ and η are small find the amplitude of x in a self-excited oscillation.

8. Prove that

$$\ddot{x} - \epsilon(1 - x^4)\dot{x} + x = 0$$

has a stable limit cycle. Find its amplitude when ϵ is small and estimate its period when ϵ is large.

9. *Rayleigh's equation* is

$$\ddot{z} + \epsilon(\tfrac{1}{3}\dot{z}^2 - \dot{z}) + z = 0.$$

If $\epsilon << 1$ show that the amplitude of the self-excited oscillation is 2. By making the change of variables $z = \epsilon w$, $t = \epsilon\tau$ discuss the relaxation oscillations by the second method of section 5.5 and estimate their period. What information can be deduced about van der Pol's equation?

10. An E.M.F. $E \sin \omega t$ is connected in series with inductance L, resistance R and a non-linear capacitor. The potential difference across the capacitor is $(Q + \beta Q^3)/C$ when its charge is Q. Show that, with $t = \tau\sqrt{LC}$,

$$Q'' + DQ' + Q + \beta Q^3 = E_0 \sin \omega\tau\sqrt{LC}$$

where $D = R\surd(C/L)$, $E_0 = EC$ and primes indicate derivatives with respect to τ. This is *Duffing's equation* with damping.

If $\omega\sqrt{LC} = 1 + \delta$ and D, β, E_0 and δ are all small, find the equation determining A so that $Q = A \sin\{(1+\delta)\,\tau + \theta\}$. Prove that, for fixed E_0, the maximum value of A is E_0/D, the corresponding value of δ being $3\beta E_0^2/8D^2$. Show also that, if $4\delta^2 > 3D^2$ and $\delta/\beta > 0$, there may be three possible values for A.

Answers to Exercises

Chapter I:

1. $\sin^3\theta = b/a$.
2. If $D\hat{A}B = 2\theta$, $\cos\theta = 0$ or $\sin\theta = 3ka/(9ka - 4mg)$. The first is stable when it is the only position and unstable when there are 3 positions. The second is always stable.
4. $45°$.
6. $2kd^2 > mgl$.

Chapter II:

1. 3. $4\pi\sqrt{\{ma/2g(2m - m_1)\}}$; 6. $2\pi l\sqrt{\{m/(2kd^2 - mg\, l)\}}$;
   2. $4\pi\sqrt{\{2ma/3(2mg - 3ka)\}}$.
2. $2\pi/\sqrt{\left(\frac{g}{l} + \frac{k}{m}\right)}$.
3. $8\pi\sqrt{\left\{5/\left(\frac{8g}{l} + \frac{k}{m}\right)\right\}}$.
7. 13 secs.
10. $4{\cdot}3''$.
11. $9{\cdot}3°$.
13. $k < 90\, m$.
14. $4{\cdot}3 \times 10^{-4} M$ radians.

Chapter III:

1. 4·4 c.p.s.; 1·7 poundals-sec./ft.; 0·0032.
2. $(9mk - 4D^2)^{\frac{1}{2}}/2\pi m$.
3. 3 secs.
5. $\epsilon X_1/X$.
7. $\dfrac{ma^2v^2(d^2k^2 + v^2D^2)^{\frac{1}{2}}}{d\{(d^2k^2 - mv^2)^2 + D^2d^2v^2\}^{\frac{1}{2}}}$. Damping reduces effect of resonance and removes transients which would otherwise be present even if the road became smooth.
8. $\omega^2/\Omega^2 = \{(1 + 8b^2)^{\frac{1}{2}} - 1\}/4b^2$.
9. (a) <205 r.p.m. (b) >427 r.p.m.

Chapter IV:

1. $e^{-400t}(\cos 1960t + 0{\cdot}20 \sin 1960t) \times 10^{-4}$ coulombs.
2. 10^{-6} farads.
7. $\dfrac{C_1\omega\,\Omega^2E_0}{C_2(\Omega^2 - \omega^2)}\left\{\dfrac{1 - \omega^2LC_2}{\omega}\sin\omega t - \dfrac{1 - \Omega^2LC_2}{\Omega}\sin\Omega t\right\}$ where $\Omega^2 = 1/L(C_1 + C_2)$.

8. $\dfrac{E_0}{102}(27 \cos 4t - 45 \sin 4t - 10e^{-t} + 85e^{-4t})$.

9. $10\displaystyle\int_0^t e^{-5T} \cos 5\tau\, E(t-\tau)d\tau$; (i) $t - \dfrac{1}{5}e^{-5t} \sin 5t$;

(ii) $\dfrac{1}{5}(3 \sin 10t - 4 \cos 10t) + \dfrac{2}{5}e^{-5t}(2 \cos 5t - \sin 5t)$.

Chapter V:

1. (i) The system moves to or from equilibrium according as $k \lesseqgtr 0$. (ii) Depending on the initial conditions either the system moves to equilibrium with $x<0$ or x increases indefinitely. (iii) The origin is a node and the system tends to equilibrium. (iv) The origin is a focus ($\rho = Ae^{-\frac{1}{3}\psi}$) and the system spirals away from it.
2. The trajectories are $\frac{1}{2}L\dot{Q}^2 + Q^2 + a\,Q^3 + b\,Q^4 = \text{constant}$ so that $Q=0$ is a centre.
3. If $\omega^2 < ag$ there is periodic motion about the vertex of the parabola which is a centre. If $\omega^2 = ag$ the particle is at rest or moves steadily in one direction. If $\omega^2 > ag$ the vertex is a saddle-point and the particle either moves steadily in one direction or approaches the vertex and then returns without reaching it.
4. If $0<\mu<1$ there are centres at $\theta = \pm \cos^{-1}\mu$ and saddle-points at $\theta = 0, \pm\pi$. If $\mu > 1$, there is a centre at $\theta = 0$ and saddle-points at $\theta = \pm\pi$.
5. The trajectories are $2\lambda^2\dot{Q}^2 = 2C\lambda^2 e^{-2\lambda Q} + 1 - 2\lambda Q$ where $\lambda = R/L$ and form closed curves for $-1 < 2\lambda^2 C < 0$.
7. 2.
8. Theorem of section 5.4 shows there is a stable limit cycle. $A = 2^{\frac{3}{4}} = 1{\cdot}68$ ($\epsilon << 1$). For $\epsilon >> 1$ limits of x are $\pm 1{\cdot}66$ and period $2{\cdot}2\epsilon$ approximately.
9. $1{\cdot}6\epsilon$; $x = \dot{z}$ gives van der Pol's equation.
10. $(\frac{3}{4}\beta A^3 - 2\delta A)^2 + D^2A^2 = E_0^2$.

Index